SCHAUM'S OUTLINE OF

THEORY AND PROBLEMS

OF

REAL VARIABLES

LEBESGUE MEASURE and INTEGRATION

with Applications to Fourier Series

•

BY

MURRAY R. SPIEGEL, Ph.D.

Professor of Mathematics
Rensselaer Polytechnic Institute

•

SCHAUM'S OUTLINE SERIES

McGRAW-HILL BOOK COMPANY

New York, St. Louis, San Francisco, London, Sydney, Toronto, Mexico, and Panama

60221

1 2 3 4 5 6 7 8 9 0   MHUN   7 2 1 0 6 9

Typography by Jack Margolin, Signs & Symbols Inc., New York

Preface

One of the most important contributions to modern mathematical analysis is the theory of measure and integration developed by Henri Lebesgue during the early years of the 20th century. Lebesgue measure and integration has many advantages over ordinary Riemann integration from the point of view of application as well as theory. It is an indispensable part of the foundations of various fields as, for example, the theory of probability and statistics and Fourier series.

In recent years Lebesgue theory has become an essential part of the traditional course in the theory of functions of a real variable, also called, for brevity, real variables or real analysis. It is the purpose of this book to present the fundamentals of Lebesgue measure and integration together with those important aspects of real variable theory needed for its understanding.

The book has been designed as a supplement to all current standard textbooks or as a textbook for a formal course in real variables. It should also prove useful in other courses in mathematics which require a knowledge of Lebesgue theory. In addition it should be of value to those readers in science and engineering, as well as mathematics, who desire an introduction to this important theory.

Each chapter begins with a clear statement of pertinent definitions, principles and theorems together with illustrative and other descriptive material. This is followed by graded sets of solved and supplementary problems. The solved problems serve to illustrate and amplify the theory, bring into sharp focus those fine points without which the student continually feels himself on unsafe ground, and provide the repetition of basic principles so vital to effective learning. Numerous proofs of theorems and derivations of basic results are included among the solved problems. The large number of supplementary problems serve as a review and possible extension of the material of each chapter.

Topics covered include the Lebesgue theory of measure, measurable functions, the Lebesgue integral and its properties, differentiation and integration. Added features are the chapters on mean convergence and applications to Fourier series including the important Riesz-Fischer theorem, as well as three appendices on the Riemann integral, summability of Fourier series, and double Lebesgue integrals and Fubini's theorem. The first chapter gives the fundamental concepts of real variable theory involving sets, functions, continuity, etc., and may either be read at the beginning or referred to as needed, according to the background of the student.

Considerably more material has been included here than can be covered in most courses. This has been done to make the book more flexible, to provide a more useful book of reference and to stimulate further interest in the topics.

I wish to take this opportunity to thank Daniel Schaum, Nicola Monti and Henry Hayden for their splendid editorial cooperation.

M. R. SPIEGEL

Rensselaer Polytechnic Institute
October 1969

CONTENTS

CONTENTS

Chapter 1

Fundamental Concepts

SETS

Fundamental in mathematics is the concept of a *set* which can be thought of as a collection of objects called *members* or *elements* of the set. In general, unless otherwise specified, we denote a set by a capital letter such as A, B, X, S, etc., and an element by a lower case letter such as a, b, x, etc. If an element a belongs to a set S we write $a \in S$. If a does not belong to S we write $a \notin S$. If a and b belong to S we write $a, b \in S$. Synonyms for set are *class, aggregate* and *collection*.

A set can be described by actually listing its elements in braces separated by commas or, if this is not possible, by describing some property held by all elements. The first is sometimes called the *roster method* while the second is called the *property method*.

    Example 1.   The set of all vowels in the English alphabet can be described by the roster method as $\{a, e, i, o, u\}$ or by the property method as $\{x : x$ is a vowel$\}$ read "the set of all elements x such that x is a vowel". Note that the colon : is read "such that".

    Example 2.   The set $\{x : x$ is a triangle in a plane$\}$ is the set of all triangles in a plane. Note that the roster method cannot be used here.

SUBSETS

If each element of a set A also belongs to a set B we call A a *subset* of B, written $A \subset B$ or $B \supset A$, and read "A is contained in B" or "B contains A" respectively. If $A \subset B$ and $B \subset A$ we call A and B *equal* and write $A = B$.

If A is not equal to B we write $A \neq B$.

If $A \subset B$ but $A \neq B$, we call A a *proper subset* of B.

It is clear that for all sets A we have $A \subset A$.

    Example 3. $\{a, i, u\}$ is a proper subset of $\{a, e, i, o, u\}$.

    Example 4. $\{i, o, a, e, u\}$ is a subset but not a proper subset of $\{a, e, i, o, u\}$ and in fact the two sets are equal. Note that a mere rearrangement of elements does not change the set.

The following theorem is true for any sets A, B, C.

Theorem 1-1.   If $A \subset B$ and $B \subset C$, then $A \subset C$.

UNIVERSAL SET AND EMPTY SET

For many purposes we restrict our discussion to subsets of some particular set called the *universe of discourse* [or briefly *universe*], *universal set* or *space* denoted by $\mathcal{U}$. The elements of a space are often called *points of the space*.

It is useful to consider a set having no elements at all. This is called the *empty set* or *null set* and is denoted by $\emptyset$. It is a subset of any set.

1

REAL NUMBERS

One of the most important sets for the purposes of this book is the set R of real numbers. It is assumed that the student is already acquainted with many properties of real numbers from the calculus. Geometric intuition is often provided by using the fact that every real number can be represented by a *point* on a line called the *real line* and conversely [see Fig. 1-1]. This enables us to speak of *sets of points* or *point sets* rather than sets of real numbers and conversely.

For example $\{x : a < x < b\}$ is an *open interval* in R and is often denoted briefly by $a < x < b$ or (a, b); $\{x : a \leqq x \leqq b\}$ is a *closed interval* in R denoted briefly by $a \leqq x \leqq b$ or $[a, b]$; $\{x : a < x \leqq b\}$ or $\{x : a \leqq x < b\}$ are *half open* [or *half closed*] *intervals* in R.

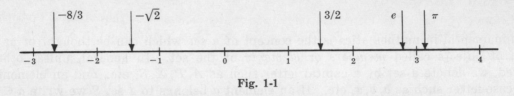

Fig. 1-1

We often find it convenient to extend the set of real numbers to include $-\infty$ and $+\infty$ or ∞. Thus $\{x : -\infty < x < \infty\}$ denoted briefly by $-\infty < x < \infty$ or $(-\infty, \infty)$ is R.

The following are important subsets of R familiar to the student.

1. **The Natural Numbers** $\{1, 2, 3, \ldots\}$ used in counting. This set is often called the set of *positive integers* and is denoted by N.

2. **The Integers** consisting of elements $0, \pm 1, \pm 2, \pm 3, \ldots$ and denoted by Z. This set is composed of the positive integers $\{1, 2, 3, \ldots\}$, the *negative integers* $\{-1, -2, -3, \ldots\}$ and *zero* $\{0\}$. Note that $N \subset Z$.

3. **The Rational Numbers** consisting of elements such as $2/3$, $-5/2$, etc., representing the quotient p/q of integers p and q, excluding division by zero. This set is denoted by Q. Note that $Z \subset Q$.

4. **The Irrational Numbers** such as $\sqrt{2}, \pi, \sqrt[3]{5}, e$ are those real numbers which are not rational numbers.

While geometric intuition does have a place in providing ideas concerning point sets on the real line, such intuition as we shall see may not always be reliable.

COMPLETENESS OR LEAST UPPER BOUND AXIOM

A real number u is called an *upper bound* of a set S of real numbers if for all $x \in S$ we have $x \leqq u$. If an upper bound p can be found such that for all upper bounds u we have $p \leqq u$, then p is called the *least upper bound* or *supremum* of S, abbreviated l.u.b. S or sup S.

The following axiom distinguishes the real numbers from any of its proper subsets [e.g. the rationals].

> **Completeness or Least Upper Bound Axiom.** If a non-empty set of real numbers has an upper bound it has a least upper bound.

We can define *lower bound* and *greatest lower bound* or *infimum* of S, abbreviated g.l.b. S or inf S, in a similar manner and can prove that if a non-empty set of real numbers has a lower bound it has a greatest lower bound.

If a set has both an upper and lower bound it is said to be *bounded*.

VENN DIAGRAMS

A universe U can be represented geometrically by the set of points inside a rectangle. In such case subsets of U [such as A and B shown shaded in Fig. 1-2] are represented by sets of points inside circles. Such diagrams, called *Venn diagrams,* often serve to provide geometric intuition regarding possible relationships between sets.

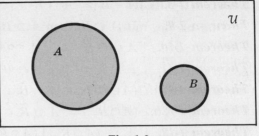

Fig. 1-2

SET OPERATIONS

1. **Union.** The set of all elements [or points] which belong to either A or B or both A and B is called the *union* of A and B and is denoted by $A \cup B$ [shaded in Fig. 1-3].

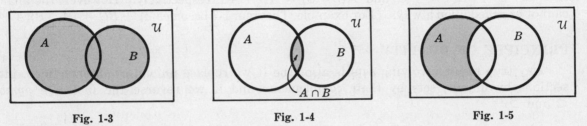

Fig. 1-3   Fig. 1-4   Fig. 1-5

2. **Intersection.** The set of all elements which belong to both A and B is called the *intersection* of A and B and is denoted by $A \cap B$ [shaded in Fig. 1-4].

   Two sets A and B such that $A \cap B = \emptyset$ i.e. which have no elements in common, are called *disjoint sets.* In Fig. 1-2, A and B are disjoint.

3. **Difference.** The set consisting of all elements of A which do not belong to B is called the *difference* of A and B denoted by $A - B$ [shaded in Fig. 1-5].

4. **Complement.** If $B \subset A$, then $A - B$ is called the *complement of B relative to A* and is denoted by $\widetilde{B}_A$ [shaded in Fig. 1-6]. If $A = U$, the universal set, we refer to $U - B$ as simply the *complement* of B and denote it by $\widetilde{B}$ [shaded in Fig. 1-7].

   The complement of $A \cup B$ is denoted by $(A \cup B)^{\sim}$.

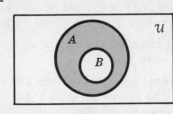 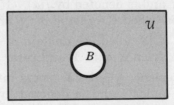

Fig. 1-6   Fig. 1-7

SOME THEOREMS INVOLVING SETS

Theorem 1-2.	$A \cup B = B \cup A$	Commutative law for unions
Theorem 1-3.	$A \cup (B \cup C) = (A \cup B) \cup C = A \cup B \cup C$	Associative law for unions
Theorem 1-4.	$A \cap B = B \cap A$	Commutative law for intersections
Theorem 1-5.	$A \cap (B \cap C) = (A \cap B) \cap C = A \cap B \cap C$	Associative law for intersections
Theorem 1-6.	$A \cap (B \cup C) = (A \cap B) \cup (A \cap C)$	First distributive law
Theorem 1-7.	$A \cup (B \cap C) = (A \cup B) \cap (A \cup C)$	Second distributive law

Theorem 1-8. $A - B = A \cap \tilde{B}$

Theorem 1-9.   If $A \subset B$, then $\tilde{A} \supset \tilde{B}$ or $\tilde{B} \subset \tilde{A}$.

Theorem 1-10. $A \cup \emptyset = A, \quad A \cap \emptyset = \emptyset$

Theorem 1-11. $A \cup \mathcal{U} = \mathcal{U}, \quad A \cap \mathcal{U} = A$

Theorem 1-12a. $(A \cup B)^{\sim} = \tilde{A} \cap \tilde{B}$ DeMorgan's first law

Theorem 1-12b. $(A \cap B)^{\sim} = \tilde{A} \cup \tilde{B}$ DeMorgan's second law

Theorem 1-13. $A = (A \cap B) \cup (A \cap \tilde{B})$ for any sets A and B.

Theorems 1-12a, 1-12b and 1-13 can be generalized [see Problems 1.71 and 1.76].

It is of interest to note that if we use the notation $A + B$ instead of $A \cup B$ and AB instead of $A \cap B$, many of the above results for an algebra of sets are reminiscent of the usual algebra of real numbers. Thus, for example, Theorems 1-3 and 1-6 become $A + (B + C) = (A + B) + C$ and $A(B + C) = AB + AC$ respectively. However, the analogy cannot be relied upon always. For example, Theorem 1-7 becomes $A + BC = (A + B)(A + C)$.

PRINCIPLE OF DUALITY

Any true result involving sets is also true if we replace unions by intersections, intersections by unions, sets by their complements and if we reverse the *inclusion symbols* $\subset$ and $\supset$.

CARTESIAN PRODUCTS

The set of all ordered pairs of elements (x, y) where $x \in A$ and $y \in B$ is called the *Cartesian product* of A and B and is denoted by $A \times B$. The Cartesian product $R \times R$ is the usual xy plane familiar from analytic geometry. In general, $A \times B \neq B \times A$.

Similarly the set of all ordered triplets (x, y, z) where $x \in A$, $y \in B$, $z \in C$ is the Cartesian product of A, B and C denoted by $A \times B \times C$.

FUNCTIONS

A *function* or *mapping* f from a set X to a set Y, often written $f : X \to Y$, is a rule which assigns to each $x \in X$ a unique element $y \in Y$. The element y is called the *image* of x under f and is denoted by $f(x)$. If $A \subset X$, then $f(A)$ is the set of all elements $f(x)$ where $x \in A$ and is called the *image of A* under f. Symbols x, y are often called *real variables*.

The set X is called the *domain* of f and $f(X)$ is called the *range* of f. If $Y = f(X)$ we say that f is from X *onto* Y and refer to f as an *onto function*.

If an element $a \in A \subset X$ maps into an element $b \in B \subset Y$, then a is called the *inverse image* of b under f and is denoted by $f^{-1}(b)$. The set of all $x \in X$ for which $f(x) \in B$ is called the *inverse image of B* under f and is denoted by $f^{-1}(B)$.

If X is a class of sets [i.e. a set whose elements are sets] then a function $f : X \to Y$ is called a *set function*.

If X and Y are sets of real numbers, f or $f(x)$ is often called a *real function*.

A function $f : X \to Y$ can also be defined as a subset of the Cartesian product $X \times Y$ such that if (x_1, y_1) and (x_2, y_2) are in this subset and $x_1 = x_2$, then $y_1 = y_2$.

ONE TO ONE FUNCTION. ONE TO ONE CORRESPONDENCE

If $f(a_1) = f(a_2)$ only when $a_1 = a_2$, we say that f is a one to one, i.e. 1-1, function.

If there exists a 1-1 function from a set X to a set Y which is both 1-1 and onto, we say that there is a *one to one, i.e. 1-1, correspondence* between X and Y.

If a function $f : X \to Y$ is *one to one* and *onto* then given any element $y \in Y$, there will be only one element $f^{-1}(y)$ in X. In such case f^{-1} will define a function from Y to X called the *inverse function*.

COUNTABILITY

Two sets A and B are called *equivalent* and we write $A \sim B$ if there exists a 1-1 correspondence between A and B.

> **Example 5.** The sets $A = \{1, 2, 3\}$ and $B = \{2, 4, 6\}$ are equivalent because of the 1-1 correspondence shown below
>
> $$A: \quad 1 \quad 2 \quad 3$$
> $$\quad \updownarrow \quad \updownarrow \quad \updownarrow$$
> $$B: \quad 2 \quad 4 \quad 6$$

If $A \sim B$, then $B \sim A$. Also if $A \sim B$ and $B \sim C$, then $A \sim C$.

A set which is equivalent to the set $\{1, 2, 3, \ldots, n\}$ for some natural number n is called *finite*; otherwise it is called *infinite*.

An infinite set which is equivalent to the set of natural numbers is called *denumerable*; otherwise it is called *non-denumerable*.

A set which is either empty, finite or denumerable is called *countable*; otherwise it is called *non-countable*.

> **Example 6.** The set Q of rational numbers is countable [see Problems 1.16 and 1.18].
>
> **Example 7.** The set of real numbers between 0 and 1 [and thus the set R] is non-countable or non-denumerable [see Problem 1.20].

The following theorems are important.

Theorem 1-14. A countable union of countable sets is countable.

Theorem 1-15 [Schroeder-Bernstein]. If $A \subset B \subset C$ and $A \sim C$, then $A \sim B$.

CARDINAL NUMBER

The *cardinal number* of the set $\{1, 2, 3, \ldots, n\}$ as well as any set equivalent to it is defined to be n. The cardinal number of any denumerable set is defined as $\aleph_0$ [*aleph null*]. The cardinal number of R, which is often called the *real continuum*, or any set equivalent to it, is defined as c or $\aleph_1$ [*aleph one*]. The cardinal number of the empty set $\emptyset$ is defined as zero (0).

Operations with infinite cardinal numbers, sometimes called *transfinite numbers*, can be defined [see Problems 1.94-1.97].

The *continuum hypothesis*, which conjectures that there is no transfinite [or cardinal] number between $\aleph_0$ and c, has never been proved or disproved. If it is proven true, then we would be justified in writing $c = \aleph_1$.

THE CANTOR SET

Consider the closed interval $[0, 1]$. Trisect the interval at points $1/3$, $2/3$ and remove the open interval $(1/3, 2/3)$ called the *middle third*. We thus obtain the set $K_1 = [0, 1/3] \cup [2/3, 1]$. By trisecting the intervals $[0, 1/3]$ and $[2/3, 1]$ and again removing the open middle thirds, we obtain $K_2 = [0, 1/9] \cup [2/9, 1/3] \cup [2/3, 7/9] \cup [8/9, 1]$. Continuing in this manner we obtain the sets $K_1, K_2, \ldots$. The Cantor set denoted by K is the intersection of $K_1, K_2, \ldots$.

It would seem that there is practically nothing left to the set K. However, it turns out [see Problem 1.22] that the set has cardinal number c, i.e. is equivalent to $[0, 1]$. It also has many other remarkable properties [see Problem 1.109 for example].

EUCLIDEAN SPACES OF n DIMENSIONS

The space defined by the Cartesian product $R^n = R \times R \times \cdots \times R$ [n times] is called n *dimensional Euclidean space* and a point in this space is an ordered n-tuplet $(x_1, x_2, \ldots, x_n)$ of real numbers. If $x = (x_1, x_2, \ldots, x_n)$, $y = (y_1, y_2, \ldots, y_n)$, the Euclidean *distance* between x and y is defined as

$$d(x, y) = \sqrt{(x_1 - y_1)^2 + (x_2 - y_2)^2 + \cdots + (x_n - y_n)^2}$$

For $n = 1$, $d(x, y) = |x - y|$ where $|a|$ denotes the *absolute value* of a [equal to a if $a \geqq 0$ and $-a$ if $a < 0$].

The set of points $\{x : d(x, y) < r\}$ is called an *open sphere* of radius r with center at y and is sometimes called a *spherical neighborhood* of y. If we replace the $<$ by $\leqq$ it is a *closed sphere* of radius r with center at y. For $n = 1$ the open or closed sphere reduces to an open or closed interval respectively.

METRIC SPACES

A *metric space* is a generalization of a Euclidean space in which there is a distance function $d(x, y)$ defined for any two elements [indicated by x, y, z] of the space satisfying the following properties.

1. $d(x, y) \geqq 0$ Non-negative property
2. $d(x, y) = d(y, x)$ Symmetric property
3. $d(x, y) = 0$ if and only if $x = y$ Zero property
4. $d(x, z) \leqq d(x, y) + d(y, z)$ Triangle inequality

SOME IMPORTANT DEFINITIONS ON POINT SETS

In this book we shall be mainly concerned with sets in the Euclidean space R of one dimension, i.e. sets of real numbers. However, although we shall adopt a language appropriate for R [e.g. distance between x and y is $|x - y|$, etc.], it should be emphasized that many of the concepts can be easily generalized to R^n or other metric spaces [by using, for example, open or closed spheres rather than intervals].

1. **Neighborhoods.** A *delta*, or δ, *neighborhood* of a point a is the set of all points x such that $|x - a| < \delta$ where δ is any given positive number. A *deleted* δ *neighborhood* of a is the set of all points x such that $0 < |x - a| < \delta$, i.e. where a itself is excluded.

   It is possible to avoid the concept of neighborhood by replacing it by an open interval [or open sphere in higher spaces].

2. **Interior Points.** A point $a \in S$ is called an *interior point* of S if there exists a δ neighborhood of a all of whose points belong to S. Alternatively, a is an interior point of S if there exists an open interval $I \subset S$ such that $a \in I$.

   The set of interior points of S is called the *interior* of S.

3. **Open Sets.** A set is said to be *open* if each of its points is an interior point.

4. **Exterior and Boundary Points.** If every δ neighborhood of a contains only points belonging to $\widetilde{S}$, then a is called an *exterior point*. If every δ neighborhood of a contains points belonging to both S and $\widetilde{S}$, then a is a *boundary point*.

The set of exterior points of S is called the *exterior* of S and the set of boundary points of S is called the *boundary* of S.

Clearly any point of a set is either an interior, exterior or boundary point.

5. **Accumulation or Limit Points.**   A point $a \in S$ is called an *accumulation point* or *limit point* of S if every deleted δ neighborhood of a contains points of S.

6. **Derived Set.**   The set of all accumulation or limit points of a set S is called the *derived set* and is denoted by S'.

7. **Closure of a Set.**   The set consisting of S together with its limit points, i.e. $S \cup S'$, is called the *closure* of S and is denoted by $\bar{S}$.

8. **Closed Sets.**   A set is called *closed* if it contains all its limit points.

9. **Open Covering of a Set.**   A class C of open intervals is said to be an *open covering* of a set S if every point of S belongs to some member of C.   If a set $J \subset C$ is an open covering of S, then we call J an *open subcovering* of S.

SOME IMPORTANT THEOREMS ON POINT SETS

Theorem 1-16.   The complement of an open set is closed and the complement of a closed set is open.

Theorem 1-17.   The union of any number of open sets is open and the intersection of a finite number of open sets is open.

Theorem 1-18.   The union of a finite number of closed sets is closed and the intersection of any number of closed sets is closed.

Theorem 1-19.   Every open set on the real line can be expressed as a countable union of disjoint open intervals [called *component intervals*] unique except as to the order of the intervals.

Theorem 1-20 [Weierstrass-Bolzano].   Every bounded infinite set in R has at least one limit point or accumulation point.

Theorem 1-21 [Heine-Borel].   Every open covering of a closed and bounded set S contains a finite open subcovering [i.e. S is covered by a *finite* number of open intervals].   For Euclidean spaces R^n this theorem is equivalent to the Weierstrass-Bolzano theorem [Theorem 1-20].   If the set is not closed and bounded the theorem is not true [see Problems 1.107 and 1.108].

COMPACT SETS

A set S is called *compact* if every open covering of S has a finite subcovering.   For R^n, compact is equivalent to closed and bounded.

LIMITS OF FUNCTIONS

A number l is said to be the *limit* of $f(x)$ as x approaches a if for every $\epsilon > 0$ there exists a $\delta > 0$ such that $|f(x) - l| < \epsilon$ whenever $0 < |x - a| < \delta$.   In such case we write $\lim_{x \to a} f(x) = l$.

Theorem 1-22.   If a limit exists it is unique.

Theorem 1-23.   If $\lim_{x \to a} f_1(x) = l_1$, $\lim_{x \to a} f_2(x) = l_2$, then

    (a) $\lim_{x \to a} [f_1(x) + f_2(x)] = l_1 + l_2$

    (b) $\lim_{x \to a} f_1(x)f_2(x) = l_1 l_2$

    (c) $\lim_{x \to a} [f_1(x)/f_2(x)] = l_1/l_2$ if $l_2 \neq 0$.

In the above definition of limit, if $0 < |x - a| < \delta$ is replaced by $0 < x - a < \delta$ then l is called the *right hand limit* and is written $\lim_{x \to a+} f(x) = l$. An analogous definition for the *left hand limit*, $\lim_{x \to a-} f(x)$, can be given. A limit will exist if and only if the left and right hand limits are equal.

CONTINUOUS FUNCTIONS

Definition: A function $f : R \to R$ is said to be *continuous* at a point a if given any $\epsilon > 0$ there exists $\delta > 0$ such that $|f(x) - f(a)| < \epsilon$ whenever $|x - a| < \delta$. Equivalently we can say that f [or $f(x)$] is continuous at a if (1) $\lim_{x \to a} f(x)$ exists, (2) $f(a)$ exists and (3) $\lim_{x \to a} f(x) = f(a)$.

We call f a *continuous function* on a set S if f is continuous at every point of S.

The following alternative definition of continuity [see Problem 1.41] is sometimes useful.

Definition: A function $f : R \to R$ is continuous at a if given any open set A containing $f(a)$, there exists an open set B containing a such that $f(B) \subset A$.

THEOREMS ON CONTINUOUS FUNCTIONS

Theorem 1-24. The sum, difference, product and quotient of continuous functions is continuous so long as division by zero is excluded.

Theorem 1-25 [Intermediate-value]. If f is continuous in $[a, b]$, then it takes on every value between $f(a)$ and $f(b)$. In particular it takes on its maximum and minimum values in $[a, b]$.

Theorem 1-26. If f is continuous on a closed set S then f is *bounded*, i.e. there exists a real number M such that $|f(x)| < M$ for all $x \in S$.

Theorem 1-27. A function f is continuous if and only if the inverse image of any open set is also open.

UNIFORM CONTINUITY

A function $f : R \to R$ is said to be *uniformly continuous* on S if given any $\epsilon > 0$ there exists a number $\delta > 0$ such that $|f(x) - f(y)| < \epsilon$ whenever $|x - y| < \delta$ where $x \in S$, $y \in S$.

Theorem 1-28. If f is continuous on a closed bounded set, it is uniformly continuous on S.

SEQUENCES

A *sequence* is a function whose domain is the set of natural numbers. It is denoted by $a_1, a_2, a_3, \ldots$ or briefly $\langle a_n \rangle$.

A sequence of real numbers $\langle a_n \rangle$ is said to *converge* to a or to have *limit* a if given any $\epsilon > 0$ there exists a positive integer n_0 such that $|a_n - a| < \epsilon$ whenever $n > n_0$. If a limit exists it is unique and the sequence is said to be *convergent* to this limit, denoted by $\lim_{n \to \infty} a_n = a$ or briefly $\lim a_n = a$.

Theorem 1-29. If $\lim a_n = a$, $\lim b_n = b$, then $\lim (a_n + b_n) = a + b$, $\lim a_n b_n = ab$, $\lim a_n/b_n = a/b$ if $b \neq 0$.

Theorem 1-30. If $\langle a_n \rangle$ is bounded [i.e. there is a constant M such that $|a_n| \leq M$] and if $\langle a_n \rangle$ is *monotonic increasing* or *monotonic decreasing* [i.e. $a_{n+1} \geq a_n$ or $a_{n+1} \leq a_n$], then $\langle a_n \rangle$ is convergent.

Theorem 1-31. A convergent sequence is bounded.

If a sequence is not convergent it is called *divergent*.

LIMIT SUPERIOR AND LIMIT INFERIOR

A number $\bar{l}$ is called the *limit superior, greatest limit* or *upper limit* of a sequence $\langle a_n \rangle$ if infinitely many terms of the sequence are greater than $\bar{l} - \epsilon$ while only a finite number of terms are greater than $\bar{l} + \epsilon$ for any $\epsilon > 0$. We denote the limit superior of $\langle a_n \rangle$ by $\limsup a_n$ or $\overline{\lim}\, a_n$.

A number $\underline{l}$ is called the *limit inferior, least limit* or *lower limit* of a sequence $\langle a_n \rangle$ if infinitely many terms of the sequence are less than $\underline{l} + \epsilon$ while only a finite number of terms are less than $\underline{l} - \epsilon$ for any $\epsilon > 0$. We denote the limit inferior of $\langle a_n \rangle$ by $\liminf a_n$ or $\underline{\lim}\, a_n$.

If infinitely many terms of $\langle a_n \rangle$ exceed any positive number M, we write $\overline{\lim}\, a_n = \infty$. If infinitely many terms are less than $-M$ where M is any positive number, we write $\underline{\lim}\, a_n = -\infty$.

Theorem 1-32.   Every bounded sequence always has a finite lim sup [or $\overline{\lim}$] and lim inf [or $\underline{\lim}$] and the sequence converges if the two are equal.

NESTED INTERVALS

Theorem 1-33.   Let $I_1 = [a_1, b_1]$, $I_2 = [a_2, b_2]$, ..., $I_n = [a_n, b_n]$, ... be a sequence of intervals such that $I_1 \supset I_2 \supset \cdots \supset I_n \supset \cdots$. Then if $\lim_{n \to \infty} (a_n - b_n) = 0$ there is one and only one point common to all the intervals. The intervals in such cases are called *nested intervals*.

CAUCHY SEQUENCES

The sequence $\langle a_n \rangle$ of real numbers is said to be a *Cauchy sequence* if given any $\epsilon > 0$ there exists a positive integer n_0 such that $|a_p - a_q| < \epsilon$ whenever $p > n_0$, $q > n_0$.

Theorem 1-34.   Every convergent sequence is a Cauchy sequence.

COMPLETENESS

A set S of real numbers is said to be *complete* if every Cauchy sequence has a limit belonging to S.

Theorem 1-35.   Every Cauchy sequence of real numbers is convergent.

> **Example 8.**   The set Q of rational numbers is not complete, but by Theorem 1-35 the set R of real numbers is complete.

SEQUENCES OF FUNCTIONS. UNIFORM CONVERGENCE

Let $f_1, f_2, \ldots$ denoted by $\langle f_n \rangle$ be a sequence of functions from A to B where $A, B \in R$. We say that $\langle f_n \rangle$ *converges uniformly* to some function f in A if given $\epsilon > 0$ there exists a positive integer n_0 such that $|f_n(x) - f(x)| < \epsilon$ for all $n > n_0$ and all $x \in A$.

Theorem 1-36.   If $\langle f_n \rangle$ is a sequence of functions which are continuous in A and uniformly convergent to f in A, then f is continuous in A.

We shall sometimes write the sequence as $\langle f_n(x) \rangle$ instead of $\langle f_n \rangle$.

SERIES

Let $\langle a_n \rangle$ be a given sequence and consider a new sequence $\langle s_n \rangle$ where

$$s_1 = a_1, \quad s_2 = a_1 + a_2, \quad s_3 = a_1 + a_2 + a_3, \quad \ldots$$

We shall call the sequence $\langle s_n \rangle$ an *infinite series* and denote it by $\sum_{n=1}^{\infty} a_n = a_1 + a_2 + \cdots$ or briefly by $\sum a_n$. The sums $s_1, s_2, \ldots$ are called the *partial sums* of the series and $\langle s_n \rangle$ is called the *sequence of partial sums*.

The infinite series is called *convergent* or *divergent* according as the sequence of partial sums $\langle s_n \rangle$ is convergent or divergent. If $\langle s_n \rangle$ converges to s then s is called the *sum* of the infinite series.

A series $\sum a_n$ is called *absolutely convergent* if $\sum |a_n|$ converges. In such series the terms can be rearranged without altering the sum of the series. We have

Theorem 1-37. If a series is absolutely convergent it is convergent.

Uniformly convergent series of functions can be defined in terms of uniformly convergent sequences of partial sums [see Problem 1.53].

Theorem 1-38 [Weierstrass M test]. If $|f_n(x)| \leq M_n$, $n = 1, 2, 3, \ldots$ where M_n are positive constants such that $\sum M_n$ converges, then $\sum f_n(x)$ is uniformly and absolutely convergent.

Solved Problems

SETS AND REAL NUMBERS

1.1. Let S be the set of all real numbers whose squares are equal to 25. Show how to describe S by (*a*) the property method and (*b*) the roster method.

(*a*) $S = \{x : x^2 = 25\}$ which is read "the set of all elements x such that $x^2 = 25$".

(*b*) Since $x^2 = 25$ for $x = 5$ and $x = -5$, we can write $S = \{5, -5\}$, i.e. S is described by actually giving its elements.

1.2. Let $A = \{x : x \text{ is an odd integer}\}$, $B = \{x : x^2 - 8x + 15 = 0\}$. Show that $B \subset A$.

Since $x^2 - 8x + 15 = 0$ or $(x-3)(x-5) = 0$ if and only if $x = 3$ or $x = 5$, we have $B = \{3, 5\}$. Then since the elements 3 and 5 are both odd integers, they belong to A. Thus every element of B belongs to A and so $B \subset A$, i.e. B is a subset of A.

1.3. Is it true that $\{2\} = 2$?

No, 2 is a *real number* while $\{2\}$ is a *set* which consists of the real number 2. A set such as $\{2\}$ consisting of only one element is sometimes called a *singleton set* and must be distinguished from the element which it contains.

1.4. Determine which of the following statements are true and correct any which are false.
(*a*) $\{x : x \neq x\} = \{\emptyset\}$. (*b*) If $A = \{x : x^2 = 4, \ x > 9\}$ and $B = \{x : x \leq 1\}$, then $B \supset A$.

(*a*) The statement is false. Any particular object is assumed to be the same as, i.e. equal to, itself. Thus there is no object which is not equal to itself. Then $\{x : x \neq x\} = \emptyset$, the empty set.

The error lies in writing $\{\emptyset\}$ rather than $\emptyset$, since $\{\emptyset\}$ is a *non-empty set* which consists of the empty set.

(*b*) Note that this is read "A is the set of all x such that $x^2 = 4$ *and* $x > 9$".

Since there is no number x such that $x^2 = 4$ [or $x = 2, -2$] and $x > 9$, it follows that $A = \emptyset$. Also since the empty set is a subset of every set, it follows that $A \subset B$ or $B \supset A$ and the statement is true.

1.5.   Prove that if $A \subset B$ and $B \subset C$, then $A \subset C$.

Let x be any element of A, i.e. $x \in A$. Then since $A \subset B$, i.e. every element of A is in B, we have $x \in B$. Also since $B \subset C$, we have $x \in C$. Thus every element of A is an element of C and so $A \subset C$.

1.6.   Let $S = \{\frac{1}{2}, \frac{2}{3}, \frac{3}{4}, \frac{4}{5}, \ldots\}$. Find (a) an upper bound, (b) a least upper bound, (c) a lower bound and (d) a greatest lower bound for S.

(a) Since all elements of S are less than 2 [for example] we can say that 2 is an upper bound. Actually any number larger than 2 is also an upper bound and there are some numbers less than 2 [e.g. $1\frac{1}{2}$, $1\frac{1}{4}$, 1] which are upper bounds.

(b) The number 1 is an upper bound of the set S and in addition it is the *least* of all upper bounds. It follows that 1 is the *least upper bound* (l.u.b.), or supremum (sup), of S. Note that 1 is not an element of S.

(c) Since all elements of S are greater than 0 [for example] we can say that 0 is a lower bound. Actually any number less than 0 is also a lower bound [e.g. $-1, -3$] and there are some numbers greater than 0 [e.g. $\frac{1}{4}, \frac{1}{2}$] which are lower bounds.

(d) The number $\frac{1}{2}$ is a lower bound of the set S and in addition it is the *greatest* of all lower bounds. It follows that $\frac{1}{2}$ is the greatest lower bound (g.l.b.), or infimum (inf), of S. Note that $\frac{1}{2}$ is an element of S.

SET OPERATIONS, VENN DIAGRAMS AND THEOREMS ON SETS

1.7.   If the universe $\mathcal{U} = \{\frac{1}{2}, 0, \pi, 5, -\sqrt{2}, -4\}$ and subsets of $\mathcal{U}$ are given by $A = \{-\sqrt{2}, \pi, 0\}$, $B = \{5, \frac{1}{2}, -\sqrt{2}, -4\}$ and $C = \{\frac{1}{2}, -4\}$, find (a) $A \cap B$, (b) $A \cup B$, (c) $(A \cup B) \cap C$, (d) $\tilde{B} \cup \tilde{C}$, (e) $A - B$, (f) $(B \cap C)^\sim$, (g) $(A \cap C) \cap (B \cup C)$.

(a) $A \cap B = \{-\sqrt{2}, \pi, 0\} \cap \{5, \frac{1}{2}, -\sqrt{2}, -4\} = \{-\sqrt{2}\}$

(b) $A \cup B = \{-\sqrt{2}, \pi, 0\} \cup \{5, \frac{1}{2}, -\sqrt{2}, -4\} = \{-\sqrt{2}, \pi, 0, 5, \frac{1}{2}, -4\}$

(c) $(A \cup B) \cap C = \{-\sqrt{2}, \pi, 0, 5, \frac{1}{2}, -4\} \cap \{\frac{1}{2}, -4\} = \{\frac{1}{2}, -4\}$ using the result of part (a).

(d) $\tilde{B} = $ set of all elements in $\mathcal{U}$ which are not in $B = \{0, \pi\}$.

$\tilde{C} = $ set of all elements in $\mathcal{U}$ which are not in $C = \{0, \pi, 5, -\sqrt{2}\}$.

Then $\tilde{B} \cup \tilde{C} = \{0, \pi\} \cup \{0, \pi, 5, -\sqrt{2}\} = \{0, \pi, 5, -\sqrt{2}\}$.

(e) $A - B = $ set of elements in A which are not in $B = \{0, \pi\}$.

Another method.   By Theorem 1-8, page 4, we have

$$A - B = A \cap \tilde{B} = \{\tfrac{1}{2}, 0, \pi, 5, -\sqrt{2}, -4\} \cap \{0, \pi\} = \{0, \pi\}$$

(f) $B \cap C = \{5, \frac{1}{2}, -\sqrt{2}, -4\} \cap \{\frac{1}{2}, -4\} = \{\frac{1}{2}, -4\}$.

Then $(B \cap C)^\sim = \{0, \pi, 5, -\sqrt{2}\}$.

Note that this result together with that of part (d) illustrates De Morgan's second law, Theorem 1-12b, page 4.

(g) $A \cap C = \{-\sqrt{2}, \pi, 0\} \cap \{\frac{1}{2}, -4\} = \emptyset$, the empty set.

$B \cap C = \{\frac{1}{2}, -4\}$ [see part (f)].

Then $\qquad\qquad (A \cap C) \cup (B \cap C) = \emptyset \cup \{\tfrac{1}{2}, -4\} = \{\tfrac{1}{2}, -4\}$

1.8.   (a) Prove De Morgan's first law, Theorem 1-12a, page 4: $(A \cup B)^\sim = \tilde{A} \cap \tilde{B}$. (b) Illustrate the result of part (a) by using a Venn diagram.

(a) We have

$$(A \cup B)^\sim = \{x : x \notin A \cup B\} = \{x : x \notin A, x \notin B\} = \{x : x \in \tilde{A}, x \in \tilde{B}\} = \tilde{A} \cap \tilde{B}$$

The result can be extended to any finite number of sets [see Problem 1.71].

(b) In the Venn diagram of Fig. 1-8 the shaded part represents $(A \cup B)^{\sim}$.

In Fig. 1-9, $\widetilde{A}$ is indicated by parallel lines constructed from upper right to lower left while $\widetilde{B}$ is indicated by parallel lines constructed from upper left to lower right. Then the region $\widetilde{A} \cap \widetilde{B}$ is represented by the region where both sets of lines are present, and it is seen that this is the same as the shaded region of Fig. 1-8.

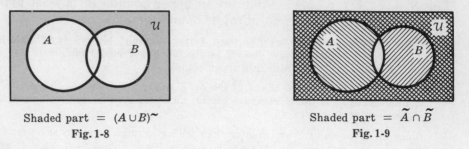

Shaded part $= (A \cup B)^{\sim}$ Shaded part $= \widetilde{A} \cap \widetilde{B}$

Fig. 1-8   Fig. 1-9

Note that a Venn diagram does not provide a proof such as is given in part (a). However, it does serve to provide possible relationships among sets which can then be proved by methods similar to that given in (a).

1.9. Prove the first distributive law, Theorem 1-6, page 3: $A \cap (B \cup C) = (A \cap B) \cup (A \cap C)$.

We have
$$
\begin{aligned}
A \cap (B \cup C) &= \{x : x \in A,\ x \in B \cup C\} \\
&= \{x : x \in A,\ x \in B \text{ or } x \in C\} \\
&= \{x : x \in A,\ x \in B \text{ or } x \in A,\ x \in C\} \\
&= \{x : x \in A \cap B \text{ or } x \in A \cap C\} \\
&= (A \cap B) \cup (A \cap C)
\end{aligned}
$$

1.10. Prove that for any sets A and B we have $A = (A \cap B) \cup (A \cap \widetilde{B})$.

Method 1.
$$
A = \{x : x \in A\} = \{x : x \in A \cap B \text{ or } x \in A \cap \widetilde{B}\} = (A \cap B) \cup (A \cap \widetilde{B})
$$

Method 2.

Let $C = \widetilde{B}$ in Problem 1.9. Then
$$
\begin{aligned}
A \cap (B \cap \widetilde{B}) &= (A \cap B) \cup (A \cap \widetilde{B}) \\
A \cap \mathcal{U} &= (A \cap B) \cup (A \cap \widetilde{B}) \\
A &= (A \cap B) \cup (A \cap \widetilde{B})
\end{aligned}
$$

The result can be generalized [see Problem 1.76].

1.11. If A, B, C are the sets of Problem 1.7, find the Cartesian products (a) $A \times C$, (b) $C \times A$.

(a) $A \times C = \{(x, y) : x \in A,\ y \in C\} = \{(-\sqrt{2}, \tfrac{1}{2}), (\pi, \tfrac{1}{2}), (0, \tfrac{1}{2}), (-\sqrt{2}, -4), (\pi, -4), (0, -4)\}$

(b) $C \times A = \{(x, y) : x \in C,\ y \in A\} = \{(\tfrac{1}{2}, -\sqrt{2}), (-4, -\sqrt{2}), (\tfrac{1}{2}, \pi), (-4, \pi), (\tfrac{1}{2}, 0), (-4, 0)\}$

Note that $A \times C \neq C \times A$ in this case.

FUNCTIONS

1.12. Determine whether a function is defined from the set X to the set Y for each diagram in (a) Fig. 1-10, (b) Fig. 1-11.

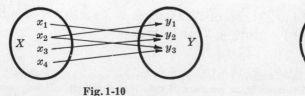 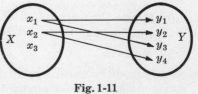

Fig. 1-10   Fig. 1-11

(a) A function is not defined from X to Y, since to the element x_2 of X there is assigned *two* distinct elements y_1 and y_3 of Y.

(b) A function is not defined from X to Y, since *no* element of Y is assigned to the element x_3 of X.

1.13. (a) Show that the diagram of Fig. 1-12 defines a function f from the set X to the set Y, i.e. $f : X \to Y$. (b) Find $f(c)$. (c) If a subset of X is given by $A = \{a, d, b\}$, find $f(A)$. (d) Find $f(X)$. (e) Find the domain of f. (f) Find the range of f. (g) Is f an onto function? Explain. (h) If $B = \{1, 2\}$, find $f^{-1}(B)$.

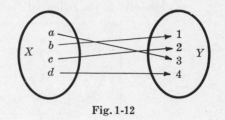

Fig. 1-12

(a) A function f is defined from X to Y since to each element of X there is a unique element of Y. The fact that no element of X happens to be assigned to the element 3 of Y or that both elements a and d of X are assigned to the same element 4 of Y, does not alter this.

(b) Since the element assigned to c is 2 we have $f(c) = 2$. Another way of saying this is that the *image* of c under f is 2.

(c) $f(A)$ is the set of all elements $f(x)$ where $x \in A$ and is given by the set $\{f(a), f(d), f(b)\} = \{4, 4, 1\} = \{4, 1\}$.

(d) $f(X) = $ the set of all elements $f(x)$ where $x \in X = \{f(a), f(b), f(c), f(d)\} = \{4, 1, 2, 4\} = \{4, 1, 2\}$.

(e) Domain of $f = X = \{a, b, c, d\}$.

(f) Range of $f = f(X) = \{4, 1, 2\}$ by part (d).

(g) Since $Y = \{1, 2, 3, 4\}$ and $f(X) = \{4, 1, 2\}$, we see that $Y \neq f(X)$. Thus f is not an onto function.

(h) $f^{-1}(B) = $ set of all $x \in X$ for which $f(x) \in B = \{b, c\}$. Note that b and c are *inverse images* of 1 and 2 respectively, i.e. $f^{-1}(b) = 1$ and $f^{-1}(c) = 2$.

1.14. Let f be a function whose value at any point x is given by $f(x) = 3x^2 - 4x + 2$ where $-1 \leq x \leq 2$. Find (a) the domain of f, (b) the range of f, (c) $f(1)$, (d) $f(3)$.

(a) Domain of f is $X = \{x : -1 \leq x \leq 2\}$ which we agree to write briefly as $-1 \leq x \leq 2$.

(b) Range of f is $Y = \{f(x) : -1 \leq x \leq 2\}$, i.e. the set of all values $f(x)$ where $-1 \leq x \leq 2$.

(c) $f(1) = 3(1)^2 - 4(1) + 2 = 1$

(d) $f(3)$ is not defined, since $x = 3$ is not in the domain of f.

1.15. Determine whether the function of Problem 1.13 is a one to one function.

A function will be one to one, or 1-1, if it assigns different images to different elements of the domain. Since this is true for the function of Problem 1.13 it is a one to one function. Note, however, that it is not an onto function [see Problem 1.13(g)].

COUNTABILITY AND CARDINAL NUMBERS

1.16. Prove that the set of rational numbers in the interval $[0, 1]$ is countably infinite or denumerable.

We must show that there is a 1-1 correspondence between the rational numbers and the natural numbers or, in other words, that the set of rational numbers in $[0, 1]$ is equivalent to the set of natural numbers.

The 1-1 correspondence is indicated as follows.

$$
\begin{array}{ccccccccccccc}
0 & 1 & \frac{1}{2} & \frac{1}{3} & \frac{2}{3} & \frac{1}{4} & \frac{3}{4} & \frac{1}{5} & \frac{2}{5} & \frac{3}{5} & \frac{4}{5} & \frac{1}{6} & \cdots \\
\updownarrow & \updownarrow & \updownarrow & \updownarrow & \updownarrow & \updownarrow & \updownarrow & \updownarrow & \updownarrow & \updownarrow & \updownarrow & \updownarrow & \\
1 & 2 & 3 & 4 & 5 & 6 & 7 & 8 & 9 & 10 & 11 & 12 & \cdots
\end{array}
$$

Note that the rational numbers are ordered according to increasing denominators. A rational number such as $\frac{2}{4}$, which is the same as $\frac{1}{2}$, is omitted since it has clearly already been counted.

1.17. Prove Theorem 1-14: A countable union of countable sets is countable.

Consider the sets

$$S_1 = \{a_{11}, a_{21}, a_{31}, \ldots\}, \quad S_2 = \{a_{12}, a_{22}, a_{32}, \ldots\}, \quad \ldots$$

There is a countable number of sets $S_1, S_2, \ldots$ and each set itself is countable. Now we can write the elements in the following form

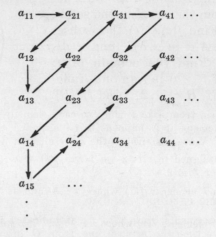

If we proceed along the directions shown we arrive at the set

$$\{a_{11}, a_{21}, a_{12}, a_{13}, a_{22}, a_{31}, a_{41}, a_{32}, \ldots\}$$

where it is seen that the first element has the sum of its subscripts equal to 2, the next two have the sum of subscripts equal to 3, the next three have sum of subscripts 4, etc. Since this establishes a 1-1 correspondence with the natural numbers, the required theorem is proved.

1.18. If Q denotes the set of rational numbers, prove that the following sets are countable: (a) $\{x : x \in Q, x \geqq 1\}$, (b) $\{x : x \in Q, x \geqq 0\}$, (c) the set Q.

(a) Let x be any rational number such that $x \geqq 1$. Then to each such x there corresponds one and only one rational number $y = 1/x$ such that $0 < y \leqq 1$. But since the set $\{y : y \in Q, 0 < y \leqq 1\}$ is countable, so also is the equivalent set $\{x : x \in Q, x \geqq 1\}$.

(b) We have
$$\{x : x \in Q, x \geqq 0\} = \{x : x \in Q, 0 \leqq x \leqq 1\} \cup \{x : x \in Q, x > 1\}$$
and the result follows by Problem 1.17 since each of the sets on the right is countable.

(c) The set $\{x : x \in Q, x > 0\}$ is countable as a consequence of part (b). Then by letting $y = -x$ we see that the equivalent set $\{x : x \in Q, x < 0\}$ is countable. Thus from Problem 1.17 and the fact that
$$Q = \{x : x \in Q, x > 0\} \cup \{x : x \in Q, x < 0\} \cup \{x : x = 0\}$$
we see that Q is countable.

1.19. Find the cardinal number of the sets in Problem 1.18.

Each of the sets is countably infinite, i.e. denumerable, and thus has the cardinal number of the set of natural numbers denoted by $\aleph_0$.

1.20. Prove that the set of all real numbers in $[0, 1]$ is non-denumerable.

Every real number in $[0, 1]$ has a decimal expansion $0.a_1 a_2 a_3 \ldots$ where $a_1, a_2, \ldots$ are any of the digits $0, 1, 2, \ldots, 9$.

We assume that numbers whose decimal expansions terminate, such as 0.7324, are written $0.73240000\ldots$ and that this is the same as $0.73239999\ldots$

If all real numbers in $[0, 1]$ are countable, then we can place them in 1-1 correspondence with the natural numbers as in the following list.

$$1 \leftrightarrow 0.a_{11}a_{12}a_{13}a_{14}\ldots$$
$$2 \leftrightarrow 0.a_{21}a_{22}a_{23}a_{24}\ldots$$
$$3 \leftrightarrow 0.a_{31}a_{32}a_{33}a_{34}\ldots$$
$$\vdots \qquad \vdots$$

We now form a number $0.b_1b_2b_3b_4$

where $b_1 = 6$ if $a_{11} = 5$ and $b_1 = 5$ if $a_{11} \neq 5$, $b_2 = 6$ if $a_{22} = 5$ and $b_2 = 5$ if $a_{22} \neq 5$, etc. [The choice 5 and 6 can of course be replaced by two other numbers]. It is then clear that the number $0.b_1b_2b_3b_4\ldots$ is different from each number in the above list and so cannot be in the list, contradicting the assumption that all numbers in $[0,1]$ were included.

Because of this contradiction it follows that the real numbers in $[0,1]$ cannot be placed in 1-1 correspondence with the natural numbers, i.e. the set of real numbers in $[0,1]$ is non-denumerable. This set has cardinal number c.

Similarly we can prove that the set of all real numbers is non-denumerable and has cardinal number c.

1.21. Prove that the sets of points on each of two line segments have the same cardinal number.

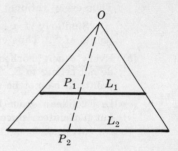

Let the two line segments be L_1 and L_2 as indicated in Fig. 1-13. By constructing the dashed line from O intersecting L_1 in point P_1 and L_2 in point P_2, we see that to each point on L_1 there corresponds one and only one point on L_2 and conversely. Thus the sets of points on L_1 and L_2 are equivalent and have the same cardinal number.

It should be noted that the ideas presented would tend to contradict the intuition of the uninitiated, since one might expect a line segment to have "more points" than another line segment which has a shorter length.

Fig. 1-13

1.22. Prove that the Cantor set [page 5] is non-denumerable.

Every number in the Cantor set has a *ternary expansion* (i.e. expansion in the scale of 3) of the form

$$\sum_{k=1}^{\infty} \frac{a_k}{3^k} \tag{1}$$

where each a_k is either 0 or 2 [see Problem 1.98]. Also each number in $[0,1]$ has a *binary expansion* (i.e. expansion in the scale of 2) of the form

$$\sum_{k=1}^{\infty} \frac{b_k}{2^k} \tag{2}$$

where each b_k is either 0 or 1 and where we assume that not all b_k after a certain term are 0 (i.e. the series is *non-terminating*).

We now set up a 1-1 correspondence between (1) and (2) such that $b_k = 0$ when $a_k = 0$ and $b_k = 1$ when $a_k = 2$. Then since the set of all real numbers in $[0,1]$ is non-denumerable [i.e. has cardinal number c] so also is the Cantor set.

DEFINITIONS INVOLVING POINT SETS

1.23. Given the point set $S = \{1, \frac{1}{2}, \frac{1}{3}, \frac{1}{4}, \ldots\}$ in R. (a) What are the interior, exterior and boundary points of S? (b) What are the accumulation or limit points of S, if any? (c) Is S open? (d) Is S closed? (e) What is the derived set S' of S? (f) Is S' closed? (g) Are the limit points of S interior, exterior or boundary points of S?

(a) Every δ neighborhood of any point $1/n$, $n = 1, 2, 3, \ldots$ contains points which belong to S and points which do not belong to S [i.e. points of $\widetilde{S}$]. Thus every point of S is a boundary point. There are no interior points of S and the exterior points of S are the points of $\widetilde{S}$, i.e. all points of R which do not belong to S.

(b) Since every deleted δ neighborhood of 0 contains points of S, 0 is a limit point. It is the only limit point of S. Note that since S is bounded and infinite, the Weierstrass-Bolzano theorem [Theorem 1-20, page 7] predicts *at least one* limit point.

(c) Since no point of S is an interior point, S is not open.

(d) S is not closed since the limit point 0 does not belong to S.

(e) $S' =$ set of all limit points of $S = \{0\}$.

(f) $\bar{S} = S \cup S' = \{1, \frac{1}{2}, \frac{1}{3}, \frac{1}{4}, \ldots\} \cup \{0\} = \{0, 1, \frac{1}{2}, \frac{1}{3}, \frac{1}{4}, \ldots\}$. Note that $\bar{S}$ is closed since it contains all its limit points.

(g) Since $S' = \{0\}$ has no limit points, it contains all of its limit points [in a vacuous sense]. Thus S' is closed.

(h) The only limit point of S is 0. Since every δ neighborhood of 0 contains points which belong to S and points which do not belong to S, it follows that 0 is a boundary point of S.

1.24. Let S be the set of irrational numbers in $[0, 1]$. (a) What are the limit points of S if any? (b) Is S open? (c) Is S compact?

(a) If a is any rational number in $[0, 1]$, then every δ neighborhood of a contains points of S. Thus every rational number in $[0, 1]$ is a limit point.

Similarly if a is any irrational number in $[0, 1]$, then every δ neighborhood of a contains points of S. Thus every irrational number in $[0, 1]$ is a limit point.

(b) Every δ neighborhood of any point $a \in S$ contains points which belong to S and points which do not belong to S. Thus every point of S is a boundary point. Then since there are no interior points, S cannot be open.

(c) In Euclidean space R^n, *compact* is the same as closed and bounded. Then since S is bounded but not closed [since it does not contain its limit points], S is not compact.

THEOREMS ON POINT SETS

1.25. Prove that if S is an open set, then the complement $\widetilde{S}$ is closed.

Let a be a limit point of $\widetilde{S}$ and suppose that $a \in S$. Then there exists an open interval (or δ neighborhood) I such that $a \in I \subset S$, and so a cannot be a limit point of $\widetilde{S}$.

This contradiction shows that $a \notin S$, i.e. $a \in \widetilde{S}$. Since a can be any limit point of $\widetilde{S}$, it follows that $\widetilde{S}$ contains all its limit points and is therefore closed.

1.26. Prove that if S is a closed set, then the complement $\widetilde{S}$ is open.

Let $a \in \widetilde{S}$. Then $a \notin S$ so that it is not a limit point of S. Thus there exists an open interval (or δ neighborhood) I such that $a \in I \subset \widetilde{S}$, and so $\widetilde{S}$ must be open.

1.27. Prove that the union of any number of open sets is open.

Suppose that the sets S_α are open and let $S = \cup S_\alpha$ [i.e. the union of the sets S_α] so that $S_\alpha \subset S$. Then if a point $a \in S$, it belongs to S_α for some $\alpha = \alpha_1$. Since S_{α_1} is open, there exists an open interval [or δ neighborhood] I containing a such that $I \subset S_{\alpha_1} \subset S$.

It follows that every point $a \in S$ is an interior point and so S is open.

1.28. Prove that the intersection of a finite number of open sets is open.

Let S_1 and S_2 be open sets and consider $S = S_1 \cap S_2$. If $a \in S$, then $a \in S_1$ and $a \in S_2$. Since S_1 and S_2 are open, there exist open intervals (or δ neighborhoods) I_1, I_2 such that $a \in I_1 \subset S_1$ and $a \in I_2 \subset S_2$. Then $a \in I_1 \cap I_2 \subset S_1 \cap S_2$. Thus there exists an open interval $I_1 \cap I_2$ contained in S and containing a, so that S is open.

The result can be extended to any finite number of open sets $S_1, S_2, \ldots, S_n$ by induction. The fact that it cannot be extended to any non-finite number of sets is indicated by Problem 1.103.

1.29. Prove that the union of a finite number of closed sets is closed.

Let $S_1, S_2, \ldots, S_n$ be closed sets so that by Theorem 1-16, page 7, $\widetilde{S}_1, \widetilde{S}_2, \ldots, \widetilde{S}_n$ are open. By De Morgan's first law generalized to any finite number of sets [see Problem 1.71],

$$\left(\bigcup_{k=1}^{n} S_k \right)^{\sim} = \bigcap_{k=1}^{n} \widetilde{S}_k$$

Then since $\bigcap_{k=1}^{n} \widetilde{S}_k$ is open by Problem 1.28, it follows that $\left(\bigcup_{k=1}^{n} S_k \right)^{\sim}$ is open so that $\bigcup_{k=1}^{n} S_k$ is closed.

1.30. Prove that the intersection of any number of closed sets is closed.

If S_α are closed sets, then $\widetilde{S}_\alpha$ are open sets. By De Morgan's second law generalized to any number of sets [see Problem 1.71],

$$(\cap S_\alpha)^{\sim} = \cup \widetilde{S}_\alpha$$

Then since $\cup \widetilde{S}_\alpha$ is open by Problem 1.27, it follows that $\cap S_\alpha$ is closed.

1.31. Let S be an open set and suppose that point $p \in S$. Prove that (a) there exists a largest open interval $I_p \subset S$ such that $p \in I_p$ and (b) $S = \cup I_p$.

(a) Since S is open, then by definition there exist open intervals in S, say (a_n, b_n), which contain p. Let $a = \text{g.l.b. } a_n$ and $b = \text{l.u.b. } b_n$ [it may happen that $a = -\infty$, $b = \infty$]. Then the interval $I_p = (a, b)$ is the largest open interval belonging to S and containing p.

(b) Every point $p \in S$ also belongs to $\cup I_p$ and every point $p \in \cup I_p$ also belongs to S. Thus $S = \cup I_p$.

1.32. Referring to Problem 1.31, let I_p, I_q be the largest open intervals of S containing points p and q respectively. (a) Prove that either $I_p = I_q$ or $I_p \cap I_q = \emptyset$. (b) Thus prove that $S = \cup I_p$ where the I_p, called *component intervals*, are disjoint.

(a) Since $p \in I_p$, we also have $p \in I_p \cup I_q$; and if we assume that neither $I_p = I_q$ nor $I_p \cap I_q = \emptyset$, it follows that $I_p \cup I_q$ is an open interval containing p which is larger than I_p, contradicting the fact that I_p is the largest such interval. Thus the required result follows.

(b) This follows at once from part (a) and Problem 1.31.

1.33. Prove that the component intervals I_p of Problem 1.32(b) are countable.

Each interval I_p contains a rational number. Then since the intervals are disjoint, they contain different rational numbers, i.e. to each interval I_p there corresponds a rational number. Since the rational numbers are countable, so also are the component intervals.

1.34. Prove that the representation of S as a countable union of disjoint open intervals, i.e. $S = \cup I_p$, is unique except insofar as the order of the intervals is concerned.

Assume that there are two different representations; say $\cup I_p^{(1)}$ and $\cup I_p^{(2)}$, so that point $p \in I_p^{(1)}$ and $p \in I_p^{(2)}$ where $I_p^{(1)} \neq I_p^{(2)}$. Then $I_p^{(1)}$ and $I_p^{(2)}$ would have to overlap or have only the point p in common, and so an endpoint of one of the intervals would belong to the other interval and thus belong to S. This, however, would contradict the fact that the intervals are open, and thus the endpoints do not belong to S.

1.35. Let $A_1, A_2, \ldots, A_n$ be n point sets on the real line. Show how to represent $\bigcup\limits_{k=1}^{n} A_k$ as $\bigcup\limits_{k=1}^{n} B_k$ where the B_k are suitably defined so as to be mutually disjoint.

Let $B_1 = A_1$, $B_2 = A_2 - A_1$, $B_3 = A_3 - (A_1 \cup A_2)$, $B_4 = A_4 - (A_1 \cup A_2 \cup A_3)$, $\ldots$, $B_n = A_n - (A_1 \cup A_2 \cup \cdots \cup A_n)$. Then we see that B_1 is the same as A_1, B_2 is the set A_2 without points of A_1 and so is disjoint to B_1, B_3 is the set A_3 without points of either A_1 or A_2 and so is disjoint to B_1 and B_2, and so on. Thus $B_1, B_2, \ldots, B_n$ are disjoint.

Further it is clear that $B_1 \cup B_2 = A_1 \cup A_2$, $B_1 \cup B_2 \cup B_3 = A_1 \cup A_2 \cup A_3$, and so on, so that $\bigcup\limits_{k=1}^{n} A_k = \bigcup\limits_{k=1}^{n} B_k$.

The result can be extended to a countable infinity of sets using mathematical induction.

LIMITS AND CONTINUITY

1.36. If (a) $f(x) = x^2$, (b) $f(x) = \begin{cases} x^2, & x \neq 2 \\ 0, & x = 2 \end{cases}$, prove that $\lim\limits_{x \to 2} f(x) = 4$.

(a) We must show that given any $\epsilon > 0$, there exists $\delta > 0$ (depending on ϵ in general) such that $|x^2 - 4| < \epsilon$ when $0 < |x - 2| < \delta$.

Choose $\delta \leqq 1$ so that $0 < |x - 2| < \delta \leqq 1$. Then

$$\begin{aligned}
|x^2 - 4| = |(x-2)(x+2)| &= |x-2|\,|x+2| < \delta\,|x+2| \\
&= \delta\,|(x-2) + 4| \\
&\leqq \delta(|x-2| + 4) \\
&< 5\delta
\end{aligned}$$

Take δ as 1 or $\epsilon/5$, whichever is smaller. Then we have $|x^2 - 4| < \epsilon$ whenever $0 < |x - 2| < \delta$ and the required result is proved.

(b) There is no difference between the proof for this case and the proof in (a), since in both cases we exclude $x = 2$.

1.37. Prove that $\lim\limits_{x \to 0} x \sin(1/x) = 0$.

We must show that given any $\epsilon > 0$, we can find $\delta > 0$ such that $|x \sin(1/x) - 0| < \epsilon$ when $0 < |x - 0| < \delta$.

If $0 < |x| < \delta$, then $|x \sin(1/x)| = |x|\,|\sin(1/x)| \leqq |x| < \delta$ since $|\sin(1/x)| \leqq 1$ for all $x \neq 0$.

Making the choice $\delta = \epsilon$, we see that $|x \sin(1/x)| < \epsilon$ when $0 < |x| < \delta$, completing the proof.

1.38. Let $f(x) = \begin{cases} \dfrac{|x-3|}{x-3}, & x \neq 3 \\ 0, & x = 3 \end{cases}$. (a) Graph the function. (b) Find $\lim\limits_{x \to 3+} f(x)$. (c) Find $\lim\limits_{x \to 3-} f(x)$. (d) Find $\lim\limits_{x \to 3} f(x)$.

(a) For $x > 3$, $\dfrac{|x-3|}{x-3} = \dfrac{x-3}{x-3} = 1$.

For $x < 3$, $\dfrac{|x-3|}{x-3} = \dfrac{-(x-3)}{x-3} = -1$.

Then the graph, shown in the adjoining Fig. 1-14, consists of the lines $y = 1$, $x > 3$; $y = -1$, $x < 3$ and the point $(3, 0)$.

(b) As $x \to 3$ from the right, $f(x) \to 1$, i.e.

$$\lim\limits_{x \to 3+} f(x) = 1$$

as seems clear from the graph. To prove this we must show that given any $\epsilon > 0$, we can find $\delta > 0$ such that $|f(x) - 1| < \epsilon$ whenever $0 < x - 1 < \delta$.

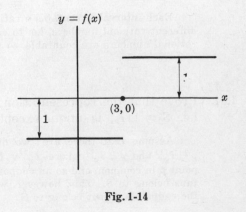

Fig. 1-14

Now since $x > 1$, $f(x) = 1$ and so the proof consists in the triviality that $|1 - 1| < \epsilon$ whenever $0 < x - 1 < \delta$.

(c) As $x \to 3$ from the left, $f(x) \to -1$, i.e. $\lim_{x \to 3-} f(x) = -1$. A proof can be formulated as in (b).

(d) Since $\lim_{x \to 3+} f(x) \neq \lim_{x \to 3-} f(x)$, $\lim_{x \to 3} f(x)$ does not exist.

1.39. Investigate the continuity of (a) $f(x) = x^2$, (b) $f(x) = \begin{cases} x^2, & x \neq 2 \\ 0, & x = 2 \end{cases}$ at $x = 2$.

(a) **Method 1.** By Problem 1.36, $\lim_{x \to 2} f(x) = 4 = f(2)$, i.e. the limit of $f(x)$ as $x \to 2$ equals the value of $f(x)$ at $x = 2$, and so f [or $f(x)$] is continuous at $x = 2$.

Method 2. We have $f(2) = 4$. Then given $\epsilon > 0$ there exists $\delta > 0$ such that $|x^2 - 4| < \epsilon$ when $|x - 2| < \delta$, by a proof similar to that in Problem 1.36. Thus $f(x)$ is continuous at $x = 2$.

(b) **Method 1.** By Problem 1.36, $\lim_{x \to 2} f(x) = 4 \neq f(2)$ [since $f(2) = 0$]. Thus the limit of $f(x)$ as $x \to 2$ is not equal to the value of $f(x)$ at $x = 2$, and so $f(x)$ is not continuous at $x = 2$.

Method 2. Given $\epsilon > 0$, we can show that there is no $\delta > 0$ such that $|f(x) - 0| = |x^2| < \epsilon$ for $|x - 2| < \delta$. Then $f(x)$ is not continuous [or is *discontinuous*] at $x = 2$.

1.40. Investigate the continuity of (a) $f(x) = x \sin(1/x)$, (b) $f(x) = \begin{cases} x \sin(1/x), & x \neq 0 \\ 5, & x = 0 \end{cases}$, (c) $f(x) = \begin{cases} x \sin(1/x), & x \neq 0 \\ 0, & x = 0 \end{cases}$ at $x = 0$.

(a) Since $f(x)$ is not defined for $x = 0$, $f(x)$ cannot be continuous at $x = 0$.

(b) By Problem 1.37, $\lim_{x \to 0} f(x) = 0 \neq f(0)$ so that $f(x)$ is not continuous at $x = 0$.

(c) By Problem 1.37, $\lim_{x \to 0} f(x) = 0 = f(0)$ so that $f(x)$ is continuous at $x = 0$.

Note that the function $f(x)$ defined in (c) is continuous in every finite interval, while the functions of (a) and (b) are continuous in every finite interval which does not include $x = 0$.

1.41. Prove that the two definitions given on page 8 for continuity at a point a are equivalent.

The result follows as a consequence of the following equivalent statements.

(1) $f(x)$ is continuous at a if given $\epsilon > 0$ there exists a $\delta > 0$ such that $|f(x) - f(a)| < \epsilon$ whenever $|x - a| < \delta$.

(2) $f(x)$ is continuous at a if given $\epsilon > 0$ there exists a $\delta > 0$ such that $f(a) - \epsilon < f(x) < f(a) + \epsilon$ whenever $a - \delta < x < a + \delta$.

(3) $f(x)$ is continuous at a if given $\epsilon > 0$ there exists a $\delta > 0$ such that $f(x) \in (f(a) - \epsilon, f(a) + \epsilon)$ whenever $x \in (a - \delta, a + \delta)$.

(4) $f(x)$ is continuous at a if $f[(a - \delta, a + \delta)]$ is contained in $(f(a) - \epsilon, f(a) + \epsilon)$.

(5) $f(x)$ is continuous at a if given any open set A containing $f(a)$ there exists an open set B containing a such that $f(B) \subset A$. [Note that $A = (f(a) - \epsilon, f(a) + \epsilon)$, $B = (a - \delta, a + \delta)$.]

1.42. Prove that $f(x) = x^2$ is uniformly continuous in $0 < x < 1$.

Method 1, using definition.

We must show that given any $\epsilon > 0$ we can find $\delta > 0$ such that $|x^2 - x_0^2| < \epsilon$ when $|x - x_0| < \delta$, where δ depends *only* on ϵ and *not* on x_0 where $0 < x_0 < 1$.

If x and x_0 are any points in $0 < x < 1$, then

$$|x^2 - x_0^2| = |x + x_0| \, |x - x_0| < |1 + 1| \, |x - x_0| = 2|x - x_0|$$

Thus if $|x - x_0| < \delta$ it follows that $|x^2 - x_0^2| < 2\delta$. Choosing $\delta = \epsilon/2$, we see that $|x^2 - x_0^2| < \epsilon$ when $|x - x_0| < \delta$, when δ depends only on ϵ and not on x_0. Hence $f(x) = x^2$ is uniformly continuous in $0 < x < 1$.

The above can also be used to prove that $f(x) = x^2$ is uniformly continuous in $0 \leqq x \leqq 1$.

Method 2:

The function $f(x) = x^2$ is continuous in the closed interval $0 \leqq x \leqq 1$. Hence by Theorem 1-28, page 8, it is uniformly continuous in $0 \leqq x \leqq 1$ and thus in $0 < x < 1$.

1.43. Prove that $f(x) = 1/x$ is not uniformly continuous in $0 < x < 1$.

Method 1:

Suppose that $f(x)$ is uniformly continuous in the given interval. Then for any $\epsilon > 0$ we should be able to find δ, say between 0 and 1, such that $|f(x) - f(x_0)| < \epsilon$ when $|x - x_0| < \delta$ for all x and x_0 in the interval.

Let $x = \delta$ and $x_0 = \dfrac{\delta}{1 + \epsilon}$. Then $|x - x_0| = \left| \delta - \dfrac{\delta}{1 + \epsilon} \right| = \dfrac{\epsilon}{1 + \epsilon} \delta < \delta$.

However, $\left| \dfrac{1}{x} - \dfrac{1}{x_0} \right| = \left| \dfrac{1}{\delta} - \dfrac{1 + \epsilon}{\delta} \right| = \dfrac{\epsilon}{\delta} > \epsilon$ (since $0 < \delta < 1$).

Thus we have a contradiction and it follows that $f(x) = 1/x$ cannot be uniformly continuous in $0 < x < 1$.

Method 2:

Let x_0 and $x_0 + \delta$ be any two points in $(0, 1)$. Then

$$|f(x_0) - f(x_0 + \delta)| = \left| \frac{1}{x_0} - \frac{1}{x_0 + \delta} \right| = \frac{\delta}{x_0(x_0 + \delta)}$$

can be made larger than any positive number by choosing x_0 sufficiently close to 0. Hence the function cannot be uniformly continuous.

SEQUENCES AND SERIES

1.44. Prove that if $\lim\limits_{n \to \infty} a_n$ exists, it must be unique.

We must show that if $\lim\limits_{n \to \infty} a_n = l_1$ and $\lim\limits_{n \to \infty} a_n = l_2$, then $l_1 = l_2$.

By hypothesis, given any $\epsilon > 0$ we can find a positive integer n_0 such that

$$|a_n - l_1| < \tfrac{1}{2}\epsilon \text{ when } n > n_0, \quad |a_n - l_2| < \tfrac{1}{2}\epsilon \text{ when } n > n_0$$

Then $\qquad |l_1 - l_2| = |l_1 - a_n + a_n - l_2| \leqq |l_1 - a_n| + |a_n - l_2| < \tfrac{1}{2}\epsilon + \tfrac{1}{2}\epsilon = \epsilon$

i.e. $|l_1 - l_2|$ is less than any positive ϵ (however small) and so must be zero. Thus $l_1 = l_2$.

1.45. If $\lim\limits_{n \to \infty} a_n = a$ and $\lim\limits_{n \to \infty} b_n = b$ prove that $\lim\limits_{n \to \infty} (a_n + b_n) = a + b$.

We must show that for any $\epsilon > 0$, we can find $n_0 > 0$ such that $|(a_n + b_n) - (a + b)| < \epsilon$ for all $n > n_0$. We have

$$|(a_n + b_n) - (a + b)| = |(a_n - a) + (b_n - b)| \leqq |a_n - a| + |b_n - b| \qquad (1)$$

By hypothesis, given $\epsilon > 0$ we can find n_1 and n_2 such that

$$|a_n - a| < \tfrac{1}{2}\epsilon \quad \text{for all } n > n_1 \qquad (2)$$

$$|b_n - b| < \tfrac{1}{2}\epsilon \quad \text{for all } n > n_2 \qquad (3)$$

Then from (1), (2) and (3),

$$|(a_n + b_n) - (a + b)| < \tfrac{1}{2}\epsilon + \tfrac{1}{2}\epsilon = \epsilon \quad \text{for all } n > n_0$$

where n_0 is chosen as the larger of n_1 and n_2. Thus the required result follows.

1.46. Prove that a convergent sequence is bounded.

Given $\lim\limits_{n\to\infty} a_n = a$, we must show that there exists a positive number M such that $|a_n| < M$ for all n. Now

$$|a_n| = |a_n - a + a| \leq |a_n - a| + |a|$$

But by hypothesis we can find n_0 such that $|a_n - a| < \epsilon$ for all $n > n_0$, i.e.,

$$|a_n| < \epsilon + |a| \quad \text{for all} \quad n > n_0$$

It follows that $|a_n| < M$ for all n if we choose M as the largest one of the numbers $a_1, a_2, \ldots, a_{n_0}$, $\epsilon + |a|$.

1.47. Find the (a) l.u.b., (b) g.l.b., (c) lim sup $(\overline{\lim})$, and (d) lim inf $(\underline{\lim})$ for the sequence $2, -2, 1, -1, 1, -1, 1, -1, \ldots$.

(a) l.u.b. $= 2$, since all terms are less than or equal to 2 while at least one term (the 1st) is greater than $2 - \epsilon$ for any $\epsilon > 0$.

(b) g.l.b. $= -2$, since all terms are greater than or equal to -2 while at least one term (the 2nd) is less than $-2 + \epsilon$ for any $\epsilon > 0$.

(c) lim sup or $\overline{\lim} = 1$, since infinitely many terms of the sequence are greater than $1 - \epsilon$ for any $\epsilon > 0$ (namely all 1's in the sequence) while only a finite number of terms are greater than $1 + \epsilon$ for any $\epsilon > 0$ (namely the 1st term).

(d) lim inf or $\underline{\lim} = -1$, since infinitely many terms of the sequence are less than $-1 + \epsilon$ for any $\epsilon > 0$ (namely all -1's in the sequence) while only a finite number of terms are less than $-1 - \epsilon$ for any $\epsilon > 0$ (namely the 2nd term).

1.48. Prove that to every set of nested intervals $[a_n, b_n]$, $n = 1, 2, 3, \ldots$, there corresponds one and only one real number.

By definition of nested intervals, $a_{n+1} \geq a_n$, $b_{n+1} \leq b_n$, $n = 1, 2, 3, \ldots$ and $\lim\limits_{n\to\infty} (a_n - b_n) = 0$.

Then $a_1 \leq a_n \leq b_n \leq b_1$, and the sequences $\{a_n\}$ and $\{b_n\}$ are bounded and respectively monotonic increasing and decreasing sequences and so converge to a and b.

To show that $a = b$ and thus prove the required result, we note that

$$b - a = (b - b_n) + (b_n - a_n) + (a_n - a) \tag{1}$$

$$|b - a| \leq |b - b_n| + |b_n - a_n| + |a_n - a| \tag{2}$$

Now given any $\epsilon > 0$, we can find n_0 such that for all $n > n_0$

$$|b - b_n| < \epsilon/3, \quad |b_n - a_n| < \epsilon/3, \quad |a_n - a| < \epsilon/3 \tag{3}$$

so that from (2), $|b - a| < \epsilon$. Since ϵ is any positive number, we must have $b - a = 0$ or $a = b$.

1.49. Prove Theorem 1-20 [Weierstrass-Bolzano]: Every bounded infinite set in R has at least one limit point.

Suppose the given bounded infinite set is contained in the finite interval $[a, b]$. Divide this interval into two equal intervals. Then at least one of these, denoted by $[a_1, b_1]$ contains infinitely many points. Dividing $[a_1, b_1]$ into two equal intervals we obtain another interval, say $[a_2, b_2]$, containing infinitely many points. Continuing this process we obtain a set of intervals $[a_n, b_n]$, $n = 1, 2, 3, \ldots$, each interval contained in the preceding one and such that

$$b_1 - a_1 = (b - a)/2, \quad b_2 - a_2 = (b_1 - a_1)/2 = (b - a)/2^2, \quad \ldots, \quad b_n - a_n = (b - a)/2^n$$

from which we see that $\lim\limits_{n\to\infty} (b_n - a_n) = 0$.

This set of nested intervals, by Problem 1.48 corresponds to a real number which represents a limit point and so proves the theorem.

1.50. Prove Theorem 1-21 [Heine-Borel]: Every open covering of a closed and bounded set S contains a finite subcovering.

Since S is closed and bounded, we can suppose that it is contained in $I = [a, b]$. Let us assume now that S does not have a finite subcovering and arrive at a contradiction.

If we bisect $I = [a, b]$, then at least one of the closed intervals $\left[a, \dfrac{a+b}{2}\right]$, $\left[\dfrac{a+b}{2}, b\right]$ has a subset of S which does not have a finite subcovering. Denote by $I_1 = [a_1, b_1]$ one of these intervals which does not have a finite subcovering.

By bisecting again and continuing in this manner, we obtain a sequence of nested closed intervals

$$I \supset I_1 \supset I_2 \supset \cdots$$

which have the property that none of the sets $S \cap I_k$ has a finite subcovering. Now the length of I_k, i.e. $L(I_k) = (b-a)/2^k$, approaches zero as $k \to \infty$ so that by the nested intervals property there is one and only one point p belonging to all the closed intervals I_k.

We now show that $p \in S$. To do this let x_1 be a point taken from $S \cap I_1$. Then take $x_2 \neq x_1$ from $S \cap I_2$, etc. We thus obtain a sequence $x_1, x_2, x_3, \ldots$ of distinct points belonging to S where $x_n \in I_n$. Since $L(I_n) \to 0$ as $n \to \infty$, it is clear that $\lim\limits_{n \to \infty} x_n = p$ which means that p is a limit point of S. But since S is closed, it follows that $p \in S$.

Now since S is covered by the set of open intervals, there will be one of these open intervals J such that $p \in J$. If k is large enough, $I_k = [a_k, b_k] \subset J$ and so $(S \cap I_k) \subset J$ which shows that $S \cap I_k$ is in fact covered by J, contradicting the statement that $S \cap I_k$ does not have a finite subcovering. This proves the theorem.

1.51. (*a*) Prove Theorem 1-34, page 9: Every convergent sequence is a Cauchy sequence.

(*b*) Prove Theorem 1-35, page 9: Every Cauchy sequence of real numbers is convergent, i.e. the set R is complete.

(*a*) Suppose the sequence $\langle a_n \rangle$ converges to l. Then given any $\epsilon > 0$, we can find n_0 such that

$$|a_p - l| < \epsilon/2 \text{ for all } p > n_0 \quad \text{and} \quad |a_q - l| < \epsilon/2 \text{ for all } q > n_0$$

Then for both $p > n_0$ and $q > n_0$, we have

$$|a_p - a_q| = |(a_p - l) + (l - a_q)| \leqq |a_p - l| + |l - a_q| < \epsilon/2 + \epsilon/2 = \epsilon$$

(*b*) Suppose $|a_p - a_q| < \epsilon$ for all $p, q > n_0$ and any $\epsilon > 0$. Then all the numbers $a_{n_0}, a_{n_0+1}, \ldots$ lie in a finite interval, i.e. the set is bounded and infinite. Hence by the Weierstrass-Bolzano theorem there is at least one limit point, say a.

If a is the only limit point, we have the desired proof and $\lim\limits_{n \to \infty} a_n = a$.

Suppose there are two distinct limit points, say a and b, and suppose $b > a$ (see Fig. 1-15). By definition of limit points, we have

$$|a_p - a| < (b-a)/3 \text{ for infinitely many values of } p \quad (1)$$

$$|a_q - b| < (b-a)/3 \text{ for infinitely many values of } q \quad (2)$$

Then since $b - a = (b - a_q) + (a_q - a_p) + (a_p - a)$, we have

$$|b - a| = b - a \leqq |b - a_q| + |a_p - a_q| + |a_p - a| \quad (3)$$

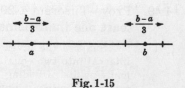

Fig. 1-15

Using (*1*) and (*2*) in (*3*), we see that $|a_p - a_q| > (b-a)/3$ for infinitely many values of p and q, thus contradicting the hypothesis that $|a_p - a_q| < \epsilon$ for $p, q > n_0$ and any $\epsilon > 0$. Hence there is only one limit point and the theorem is proved.

Note that Theorems 1-34 and 1-35 can be stated in one theorem:

A sequence of real numbers converges if and only if it is a Cauchy sequence.

1.52. Let $f_n(x) = nxe^{-nx^2}$, $n = 1, 2, 3, \ldots$, $0 \leqq x \leqq 1$. Investigate the uniform convergence of the sequence $\langle f_n(x) \rangle$.

We have $\lim\limits_{n \to \infty} f_n(x) = 0$ for $0 \leqq x \leqq 1$. To determine whether the sequence converges uniformly to 0 in this interval, we must show that given $\epsilon > 0$ there exists $n_0 > 0$ such that $|f_n(x) - 0| < \epsilon$, i.e. $|f_n(x)| < \epsilon$, for all $n > n_0$ where n_0 depends only on ϵ and not on x.

Now $|f_n(x)| = nxe^{-nx^2}$ has a maximum at $x = 1/\sqrt{2n}$ (by the usual rules of elementary calculus), the value of this maximum being $\sqrt{n/2e}$. Hence as $n \to \infty$, $f_n(x)$ cannot be made arbitrarily small for all x and so the sequence does not converge uniformly to 0.

1.53. (a) Give a definition of uniformly convergent series of functions.

(b) State and prove a theorem corresponding to Theorem 1-36, page 9, for series of functions.

(a) Let $\sum\limits_{k=1}^{\infty} f_k(x)$ be a series of functions and
$$s_n(x) = f_1(x) + f_2(x) + \cdots + f_n(x)$$
be the nth partial sum of the series. Suppose that in some set A, for example $[a, b]$, the series converges to the sum $s(x)$. We say that the series converges uniformly to $s(x)$ in $[a, b]$ if given $\epsilon > 0$ there exists a positive integer n_0 such that $|s_n(x) - s(x)| < \epsilon$ for all $n > n_0$ and all $x \in [a, b]$.

(b) The theorem which we shall prove is the following: If the functions $f_n(x)$ are continuous in a set A and if $\sum f_n(x)$ converges uniformly to $s(x)$ in A, then $s(x)$ is continuous in A. We prove the theorem for the case where $A = [a, b]$.

First we observe that if $s_n(x)$ is the nth partial sum of the series and $r_n(x)$ is the remainder of the series after n terms, then
$$s(x) = s_n(x) + r_n(x)$$
so that
$$s(x + h) = s_n(x + h) + r_n(x + h)$$
and
$$s(x + h) - s(x) = s_n(x + h) - s_n(x) + r_n(x + h) - r_n(x) \tag{1}$$
where we choose h so that both x and $x + h$ are in $[a, b]$ (if $x = b$, for example, this will require $h < 0$).

Since $s_n(x)$ is a sum of a finite number of continuous functions, it must also be continuous. Then given $\epsilon > 0$, we can find δ so that
$$|s_n(x + h) - s_n(x)| < \epsilon/3 \quad \text{whenever} \quad |h| < \delta \tag{2}$$
Since the series, by hypothesis, is uniformly convergent, we can choose n_0 depending on ϵ but not on x so that
$$|r_n(x)| < \epsilon/3 \quad \text{and} \quad |r_n(x + h)| < \epsilon/3 \quad \text{for} \quad n > n_0 \tag{3}$$
Then from (1), (2) and (3),
$$|s(x + h) - s(x)| \leqq |s_n(x + h) - s_n(x)| + |r_n(x + h)| + |r_n(x)| < \epsilon$$
for $|h| < \delta$, and so the continuity is established.

1.54. Prove the *Weierstrass M test*, [Theorem 1-38, page 10]: If $|f_n(x)| \leqq M_n$, $n = 1, 2, 3, \ldots$, where M_n are positive constants such that $\sum M_n$ converges, then $\sum f_n(x)$ is (a) uniformly and (b) absolutely convergent.

(a) The remainder of the series $\sum f_n(x)$ after n terms is $r_n(x) = f_{n+1}(x) + f_{n+2}(x) + \cdots$. Now
$$|r_n(x)| = |f_{n+1}(x) + f_{n+2}(x) + \cdots| \leqq |f_{n+1}(x)| + |f_{n+2}(x)| + \cdots \leqq M_{n+1} + M_{n+2} + \cdots$$
But $M_{n+1} + M_{n+2} + \cdots$ can be made less than ϵ by choosing $n > n_0$ since $\sum M_n$ converges. Since n_0 is clearly independent of x, we have $|r_n(x)| < \epsilon$ for $n > n_0$, and the series is uniformly convergent.

(b) The absolute convergence follows from the fact that $\sum\limits_{k=n+1}^{\infty} |f_k(x)| < \sum\limits_{k=n+1}^{\infty} M_k < \epsilon$ for $n > n_0$.

1.55. Prove that $\sum\limits_{n=1}^{\infty} \dfrac{\cos nx}{n^2}$ is uniformly convergent for all x.

We have for all x,

$$\left| \frac{\cos nx}{n^2} \right| \leqq \frac{1}{n^2}$$

Then since $\sum\limits_{n=1}^{\infty} \dfrac{1}{n^2}$ converges, the series is uniformly convergent by the Weierstrass M test of Problem 1.54.

Supplementary Problems

SETS AND REAL NUMBERS

1.56. Let S be the set of all natural numbers between 5 and 15 which are even. Describe S according to (a) the roster method, (b) the property method.
  Ans. (a) $S = \{6, 8, 10, 12, 14\}$, (b) $S = \{x : x \text{ is even}, 5 < x < 15\}$

1.57. Let $A = \{x : x^2 - 3x + 2 = 0\}$, $B = \{x : x^2 \leqq 16\}$. Determine whether or not $A \subset B$.

1.58. Prove that for any set A we have $A \subset A$.

1.59. Discuss the truth or falsity of the following statements. (a) If A and B are any two sets then either $A \subset B$, $A \supset B$, or $A = B$. (b) If x and y are any two real numbers then either $x < y$, $x > y$, or $x = y$.

1.60. Prove that any subset of the empty set must be the empty set.

1.61. Let $A = \{1, -\frac{1}{2}, \frac{1}{3}, -\frac{1}{4}, \frac{1}{5}, \ldots\}$. Find (a) l.u.b. A, (b) g.l.b. A.   *Ans.* (a) 1, (b) $-\frac{1}{2}$.

1.62. Give an example of a set S for which (a) the l.u.b. belongs to S but the g.l.b. does not, (b) the l.u.b. does not belong to S but the g.l.b. does, (c) both the l.u.b. and g.l.b. belong to S, (d) neither the l.u.b. nor the g.l.b. belong to S.

1.63. Give an example to show that a non-empty set of rational numbers can have an upper bound but not a least upper bound. Reconcile this with the completeness axiom on page 2.

1.64. Show that between any two different rational numbers a and b there is (a) at least one rational number, (b) at least one irrational number. How many rational numbers and irrational numbers would you expect to have between a and b? Explain.

1.65. Work Problem 1.64 when a and b are irrational numbers.

1.66. Let $\mathcal{S}$ be the set or class of all sets which are not elements of themselves. (a) Prove that if $\mathcal{S} \in \mathcal{S}$, then $\mathcal{S} \notin \mathcal{S}$. (b) Prove that if $\mathcal{S} \notin \mathcal{S}$, then $\mathcal{S} \in \mathcal{S}$. The paradox described is called *Russell's paradox* and shows that sets should not be "too large". (c) Illustrate the paradox by discussing the situation of the barber in a small town who shaves all those men and only those men who do not shave themselves. In particular attempt to answer the question "Who shaves the barber?"

SET OPERATIONS, VENN DIAGRAMS, AND THEOREMS ON SETS

1.67. Let a universe be given by $\mathcal{U} = \{1, 2, 3, 4, 5\}$ and suppose that subsets of $\mathcal{U}$ are $A = \{1, 5\}$, $B = \{2, 5, 3\}$, $C = \{4, 2\}$. Find (a) $A \cup (B \cup C)$, (b) $(A \cup B) \cup C$, (c) $A \cap (B \cup C)$, (d) $(A \cap B) \cup (A \cap C)$, (e) $\widetilde{A} \cap (\widetilde{B} \cap \widetilde{C})$, (f) $(A \cup B) - (A \cup C)$, (g) $(A \cap C)^{\sim} \cup B$, (h) $A - (\widetilde{B} \cup \widetilde{C})$.
  Ans. (a) $\{1, 2, 3, 4, 5\}$, (b) $\{1, 2, 3, 4, 5\}$, (c) $\{5\}$, (d) $\{5\}$, (e) $\varnothing$, (f) $\{3\}$, (g) $\{2, 5, 3\}$, (h) $\{5\}$.

1.68. Let $\mathcal{U}$ be the set of all non-negative integers and consider the subsets $A = \{x : x \text{ is an even integer}, 1 \leqq x < 6\}$, $B = \{x : x \text{ is a prime number}, 0 < x \leqq 4\}$. Find (a) $A \cup B$, (b) $A \cap B$, (c) $\widetilde{A} \cap \widetilde{B}$, (d) $A - B$, (e) $B - A$, (f) $(A - B) \cup (B - A)$.
  Ans. (a) $\{2, 3, 4\}$, (b) $\{2\}$, (c) $\{x : x \geqq 0, x \neq 2, 3, 4\}$, (d) $\{4\}$, (e) $\{3\}$, (f) $\{3, 4\}$

1.69.   Prove (a) Theorem 1-2, page 3, (b) Theorem 1-3, page 3, (c) Theorem 1-4, page 3.

1.70.   Prove De Morgan's second law, Theorem 1-12b, page 4, and (b) illustrate by using a Venn diagram.

1.71.   Generalize De Morgan's first and second laws to any number of sets.

1.72.   Illustrate the principle of duality by referring to the theorems on pages 3 and 4.

1.73.   Prove that $(A - B) \cup B = A$ if and only if $B \subset A$, and illustrate by using a Venn diagram.

1.74.   Prove or disprove: If $A - B = \emptyset$, then $A = B$.

1.75.   Prove that $A \cup B = (A - A \cap B) \cup (B - A \cap B)$ and illustrate by a Venn diagram.

1.76.   Generalize the result of Problem 1.10.

1.77.   Referring to Problem 1.67, find (a) $A \times B$, (b) $B \times A$, (c) $A \times B \times C$.

1.78.   Prove or disprove: $A \times (B \times C) = (A \times B) \times C$.

1.79.   Prove or disprove: $A \times (B \cup C) = (A \times B) \cup (A \times C)$.

FUNCTIONS

1.80.   Determine whether a function is defined from the set X to the set Y for the diagram of Fig. 1-16.

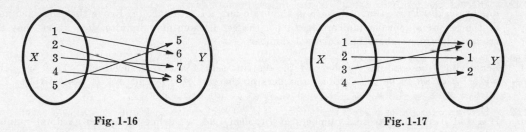

Fig. 1-16   Fig. 1-17

1.81.   (a) Show that the diagram of Fig. 1-17 defines a function from the set X to the set Y. (b) Find $f(2)$.
(c) Find $f(0)$. (d) Find the domain of f. (e) Find the range of f. (f) If A is the subset $\{2, 3\}$ of X,
find $f(A)$. (g) Find $f(X)$. (h) Is f an onto function? Explain. (i) Find the image of 1 under f.
(j) Find $f^{-1}(2)$. (k) If $B = \{0, 1\}$, find $f^{-1}(B)$. (l) Find the inverse image of 1 under f. (m) Is an
inverse function f^{-1} defined from B to A where $B = \{0, 1\}$ and $A = \{2, 3\}$?
Ans.   (b) 1   (d) $\{1, 2, 3, 4\}$ (f) $\{0, 1\}$ (h) Yes   (j) 4   (l) $\{2\}$
(c) not defined   (e) $\{0, 1, 2\}$ (g) $\{0, 1, 2\}$ (i) 0   (k) $\{1, 2, 3\}$ (m) No

1.82.   Let $f(x) = \begin{cases} x^2, & 0 < x \leqq 1 \\ 2, & x > 1 \end{cases}$ define a function f from X to Y. Find (a) the domain of f, (b) the
range of f, (c) $f(\tfrac{1}{2})$, (d) $f(3)$, (e) $f(0)$, (f) $f(-1)$, (g) $f(A)$ where $A = \{x : \tfrac{1}{2} < x < 3\}$.
Ans. (a) $x > 0$,   (b) $\{x : 0 < x \leqq 1, \ x = 2\}$,   (c) $\tfrac{1}{4}$,   (d) 2,   (e) not defined,   (f) not defined,
(g) $\{x : \tfrac{1}{4} < x \leqq 1, \ x = 2\}$

1.83.   Let a function $f : R \to R$ be defined by
$$f(x) \ = \ \begin{cases} 1 \text{ if } x \text{ is rational} \\ 0 \text{ if } x \text{ is irrational} \end{cases}$$
Find (a) $f(\tfrac{2}{3})$, (b) $f(-\sqrt{2})$, (c) $f(\pi)$, (d) $f(1.252525\ldots)$.   *Ans.* (a) 1 (b) 0 (c) 0 (d) 1

1.84.   If f is a function from X to Y, is it possible that $f(X) \supset Y$? Explain.

1.85.   Given functions $f : A \to B$ and $g : B \to C$. Then if $a \in A$, $f(a) \in B$ and $g(f(a)) \in C$. Thus
to each $a \in A$ we have $g(f(a)) \in C$ and we have a function from A to C called the *product function*
or *composition function* denoted by $g \circ f$ or gf. If for each real number x we have $f(x) = x^2 + 1$
and $g(x) = 2x$, find (a) $g(f(1))$, (b) $f(g(-2))$, (c) $g(f(x))$. How would you define $f(g(x))$? Is this
the same as $g(f(x))$? Explain.   *Ans.* (a) 4 (b) 17 (c) $2(x^2 + 1)$

1.86. Determine whether the function of Problem 1.81 is one to one.

1.87. Define a function $f : R \to R$ by $f(x) = 3x - 5$. (a) Show that the function is one to one and onto. (b) Show that the inverse function is defined by $f^{-1}(x) = \frac{1}{3}(x + 5)$.

1.88. Define the functions $f_1 : R \to R$ and $f_2 : R \to R$ by $f_1(x) = x^2 + 1$ and $f_2(x) = x^3 + 1$. (a) Explain why f_2 is 1-1 and onto while f_1 is onto but not 1-1, and (b) show that we can define an inverse function $f_2^{-1} : R \to R$ by $f_2^{-1}(x) = \sqrt[3]{x - 1}$ but cannot define an inverse function $f_1^{-1} : R \to R$.

1.89. (a) If $A = \{a, b, c, d\}$ and $B = \{1, 0\}$, determine how many functions there are from A to B. (b) Show these diagrammatically.   *Ans.* (a) $4^2 = 16$

COUNTABILITY AND CARDINAL NUMBERS

1.90. Prove that there is a one to one correspondence between the points of the interval $0 \leqq x \leqq 1$ and (a) $-4 \leqq x \leqq 6$, (b) $-4 < x < 6$. What is the cardinal number of each set?

1.91. (a) Prove that the set of all irrational numbers in $[0, 1]$ is non-denumerable and (b) find the cardinal number.

1.92. (a) Prove that the cardinal number of points inside a square is equal to the cardinal number of the set of points on one side. (b) Generalize the result to higher dimensions.

1.93. An *algebraic number* is a number which is a root of a polynomial equation $a_0 x^n + a_1 x^{n-1} + \cdots + a_n = 0$ where $a_0, a_1, \ldots, a_n$ are integers. A *transcendental number* is a number which is not algebraic. (a) Prove that $\sqrt{2} + \sqrt{3}$ is algebraic but not rational. (b) Prove that the cardinal number of the set of algebraic numbers is $\aleph_0$, i.e. the set is countably infinite. (c) Prove that the set of transcendental numbers has cardinal number $c = \aleph_1$.

1.94. Let α and β be cardinal numbers of the disjoint sets A and B respectively. Then we define $\alpha + \beta$ and $\alpha \cdot \beta$ or $\alpha\beta$ to be the cardinal numbers of the sets $A \cup B$ and $A \times B$ respectively. Prove that (a) $\aleph_0 + \aleph_0 = \aleph_0$, (b) $\aleph_0 \cdot \aleph_0 = \aleph_0$, (c) $\aleph_0 + c = c$, (d) $c \cdot c = c$.

1.95. If α and β are the cardinal numbers of Problem 1.94, we define α^β as the cardinal number of the set of all functions on B with values in A. For example if $A = (0, 1)$ and $B = (a, b, c)$, then the set of functions on B with values in A is the set of ordered triplets $(0, 0, 0), (1, 0, 0), (0, 1, 0), (0, 0, 1), (1, 1, 0), (1, 0, 1), (0, 1, 1), (1, 1, 1)$. Since this set has cardinal number 8, we see that $2^3 = 8$. Use this definition to find (a) 3^2, (b) $\aleph_0^4$. Prove that if α, β, γ are cardinal numbers, then $(\alpha^\beta)^\gamma = \alpha^{\beta\gamma}$.

1.96. (a) Prove that every real number in $[0, 1]$ can be expressed in the *scale* of 2, called a *binary expansion*, as $\sum\limits_{k=1}^{\infty} b_k / 2^k$ where the b_k are either 0 or 1.

(b) Prove that the set of all binary expansions in (a) has cardinal number $2^{\aleph_0}$.

(c) Prove that $2^{\aleph_0} = c$.

1.97. The cardinal number α of a set A is said to be greater than the cardinal number β of a set B if B is equivalent to a subset of A but A and B are not equivalent. Prove that (a) $c > \aleph_0$, (b) $2^\alpha > \alpha$. Deduce from (b) that there are infinitely many transfinite numbers.

1.98. Prove that each number of the Cantor set can be expressed as a series [called a *ternary* expansion] of the form $\sum\limits_{k=1}^{\infty} a_k / 3^k$ where the a_k are either 0 or 2.

DEFINITIONS AND THEOREMS INVOLVING POINT SETS

1.99. Let S be the set of rational numbers in $[0, 1]$. (a) What are the interior, exterior and boundary points of S? (b) What are the accumulation or limit points of S if any? (c) Is S open? (d) Is S closed? (e) What is the derived set S'? (f) What is the closure $\bar{S}$? (g) Is S' closed? (h) Are the limit points of S interior, exterior or boundary points of S?

1.100. Work Problem 1.99 if S is (a) the set of irrational numbers in $[0, 1]$ and (b) the set of all real numbers in $[0, 1]$.

1.101. Show that the set of real numbers R is both open and closed.

1.102. (*a*) If a limit point of a set does not belong to the set, prove that it must be a boundary point of the set.

(*b*) Give a definition of limit point using open intervals instead of δ neighborhoods and use this definition to give a proof of the result in (*a*).

1.103. Let $I_k = \left(-\frac{1}{k}, 1+\frac{1}{k}\right)$ where $k = 1, 2, 3, \ldots$. (*a*) Show that $\underset{k}{\cap} I_k = [0, 1]$. (*b*) Use (*a*) to prove that the intersection of a denumerable number of open sets need not be open and compare with the theorem of Problem 1.28.

1.104. Give an example showing that the union of a denumerable number of closed sets need not be closed.

1.105. Prove that $(a, b) = \overset{\infty}{\underset{k=1}{\cup}} \left[a+\frac{1}{k}, b-\frac{1}{k}\right]$ and thus show that every open interval can be expressed as a countable union of closed intervals.

1.106. Prove that every countable set can be expressed as a countable union of closed sets.

1.107. Let $I_k = (k - \frac{1}{3}, k + \frac{1}{3})$, $k = 1, 2, 3, \ldots$ be a set of open intervals which cover the set N of natural numbers $1, 2, 3, \ldots$ which is a closed set. (*a*) Is there a finite subcovering of N? (*b*) Does your answer to (*a*) contradict the Heine-Borel theorem? Explain.

1.108. Let $S = \{1, \frac{1}{2}, \frac{1}{3}, \ldots\}$ and let I_k be an interval so small that it contains only the point $1/k$. (*a*) Is there a finite subcovering of S by I_k? (*b*) Does your answer to (*a*) contradict the Heine-Borel theorem? Explain.

1.109. A set S is called *perfect* if every point of S is a limit point, i.e. $S' = S$. Prove that the Cantor set is perfect. Is the Cantor set a closed set? Explain.

LIMITS AND CONTINUITY

1.110. If $f(x) = x^2 + 3x + 5$, find $\lim_{x \to 2} f(x)$ using the definition.

1.111. Prove that (*a*) $\lim_{x \to 4} \frac{x^2 - 16}{x - 4} = 8$, (*b*) $\lim_{x \to 4} \frac{\sqrt{x} - 2}{x - 4} = \frac{1}{4}$.

1.112. Let $f(x) = \begin{cases} 3x - 1, & x < 0 \\ 0, & x = 0 \\ 2x + 5, & x > 0 \end{cases}$. Find (*a*) $\lim_{x \to 2} f(x)$, (*b*) $\lim_{x \to -3} f(x)$, (*c*) $\lim_{x \to 0+} f(x)$, (*d*) $\lim_{x \to 0-} f(x)$,

(*e*) $\lim_{x \to 0} f(x)$.   *Ans.* (*a*) 9, (*b*) −10, (*c*) 5, (*d*) −1, (*e*) does not exist.

1.113. Prove (*a*) Theorem 1-22, page 7, (*b*) Theorem 1-23, page 7, (*c*) Theorem 1-24, page 8.

1.114. Prove that $\lim_{x \to a} f(x)$ exists at $x = a$ if and only if $\lim_{x \to a+} f(x) = \lim_{x \to a-} f(x)$, i.e. the right and left hand limits are equal.

1.115. Prove that $f(x) = x^2 - 4x + 3$ is continuous at $x = 2$ by using the definition.

1.116. Prove that $f(x) = x/(x + 2)$ is continuous at all points of the interval (*a*) $1 < x < 3$, (*b*) $-2 < x \leqq 0$.

1.117. Prove that the sum of two or more continuous functions is continuous.

1.118. Prove that the (*a*) product and (*b*) quotient of two continuous functions is continuous provided that division by zero is excluded.

1.119. Prove that a polynomial is continuous in every finite interval.

1.120. Prove (*a*) Theorem 1-25, page 8, (*b*) Theorem 1-26, page 8.

1.121. Prove Theorem 1-27, page 8.

1.122. Prove that $f(x) = x^2 - 4x + 3$ is uniformly continuous in the interval $2 \leqq x \leqq 4$.

1.123. Prove that $f(x) = 1/x^2$ is (a) continuous for $x > a$ where $a \geqq 0$, (b) uniformly continuous for $x > a$ if $a > 0$, (c) not uniformly continuous in $0 < x < 1$.

1.124. Let $f(x)$ and $g(x)$ be uniformly continuous in $a \leqq x \leqq b$. Prove that (a) $f(x) + g(x)$, (b) $f(x)\,g(x)$, (c) $f(x)/g(x)$, $g(x) \neq 0$ are uniformly continuous in $a \leqq x \leqq b$.

1.125. If $f(x)$ is continuous at x_0 and $f(x_0) > 0$, prove that there exists an interval $(x_0 - h, x_0 + h)$, $h > 0$ in which $f(x) > 0$.

1.126. Prove Theorem 1-28, page 8.

SEQUENCES AND SERIES

1.127. Use the definition of limit of a sequence to prove that $\lim\limits_{n \to \infty} \dfrac{3n + 2}{4 - 5n} = -\dfrac{3}{5}$.

1.128. If $\lim\limits_{n \to \infty} b_n = b \neq 0$, prove that there exists a number n_0 such that $|b_n| > \frac{1}{2}|b|$ for all $n > n_0$.

1.129. If $\lim\limits_{n \to \infty} a_n = a$, $\lim\limits_{n \to \infty} b_n = b$, prove that (a) $\lim\limits_{n \to \infty} a_n b_n = ab$ and (b) $\lim\limits_{n \to \infty} \dfrac{a_n}{b_n} = \dfrac{a}{b}$ if $b \neq 0$.

1.130. Find lim sup ($\overline{\lim}$) and lim inf ($\underline{\lim}$) for the sequences (a) $\langle (-1)^{n+1}/n \rangle$, (b) $\langle (-1)^{n+1}(n+1)/(n+2) \rangle$, (c) $\langle (-1)^{n-1}(2n-1) \rangle$, (d) $\langle n^{[1+(-1)^n]/2} \rangle$. *Ans.* (a) $0, 0$ (b) $1, -1$ (c) $\infty, -\infty$ (d) $\infty, 1$.

1.131. (a) Prove that the limit of a sequence exists if and only if the limit inferior and limit superior of the sequence are equal. (b) Prove that if $a_n \geqq 0$ and $\overline{\lim\limits_{n \to \infty}}\, a_n = 0$ then $\lim\limits_{n \to \infty} a_n = 0$.

1.132. Prove that the sequence $\langle e^{-nx^2} \rangle$ converges uniformly to 0 in any interval not including $x = 0$.

1.133. Test for uniform convergence in $[0, 1]$ the sequence $\langle x^n \rangle$.

1.134. (a) Prove Theorem 1-37, page 10. (b) Give an example of a convergent series which is not absolutely convergent.

1.135. Test for uniform convergence the series $\sum\limits_{n=0}^{\infty} x^n$ in its interval of convergence.

Ans. Uniformly convergent for $|x| < p$ where $0 < p < 1$.

1.136. If $\sum\limits_{n=0}^{\infty} a_n x^n$ converges for $x = x_0$, prove that it converges uniformly and absolutely in the interval $|x| < |x_1|$ where $|x_1| < |x_0|$.

1.137. Test for uniform convergence: (a) $\sum\limits_{n=1}^{\infty} \dfrac{x^n}{n^2 + 1}$, (b) $\sum\limits_{n=1}^{\infty} \dfrac{1}{x^4 + n^4}$.

Ans. (a) Uniformly convergent in $-1 \leqq x \leqq 1$. (b) Uniformly convergent for all x.

1.138. Test for uniform convergence: $\sum\limits_{n=1}^{\infty} \dfrac{x^2}{(1 + x^2)^n}$.

Ans. Uniformly convergent in any interval not including $x = 0$.

1.139. Prove that the series $\dfrac{1}{1 + x^2} - \dfrac{1}{2 + x^2} + \dfrac{1}{3 + x^2} - \cdots$ is uniformly convergent but not absolutely convergent for all x.

1.140. Give an example of a series which is absolutely convergent but not uniformly convergent.

Chapter 2

Measure Theory

LENGTH OF AN INTERVAL

We define the *length* of all the intervals $a < x < b$, $a \leqq x \leqq b$, $a \leqq x < b$, $a < x \leqq b$ as $b - a$. The length of an infinite interval such as $x > a$ or $x \leqq b$ is defined to be ∞. In case $a = b$, the interval $a \leqq x \leqq b$ degenerates to a point and has length *zero*. Thus length is a non-negative real number.

Since an interval I is a set of points, we see that its length is a set function of I having value $L(I) \geqq 0$.

AREA AND VOLUME

The idea of length is easily generalized to two, three and higher dimensional Euclidean spaces. For example, if we consider two dimensions, we can think of a *rectangle* as a *generalized interval* $a < x < b$, $c < y < d$ [or some variation obtained on replacing $<$ by $\leqq$] and the *area* of the rectangle as a generalized length given by $(b-a)(d-c)$. Similarly in three dimensions we can consider *volume* as a generalized length.

Although in the following we shall restrict ourselves to intervals and lengths of intervals in one dimensional Euclidean space, i.e. the real line, it should be emphasized that generalizations to the higher dimensional spaces can be made.

LENGTH OF UNION OF DISJOINT INTERVALS. LENGTH OF EMPTY SET

We shall be interested in generalizing the notion of length to sets on the real line besides intervals. The following definitions provide some obvious extensions.

Definition 2.1. If $I_1, I_2, \ldots$ are mutually disjoint intervals, then
$$L(I_1 \cup I_2 \cup \cdots) = L(I_1) + L(I_2) + \cdots$$

Definition 2.2. $L(\emptyset) = 0$, i.e. the length of the empty set is zero.

LENGTH OF AN OPEN SET

Since any open set O can be expressed as a countable union of mutually disjoint open intervals $I_1, I_2, \ldots$ unique except insofar as order is concerned [see Theorem 1-19, page 7], we are led to the following

Definition 2.3. The length of an open set $O = \bigcup\limits_{k=1}^{\infty} I_k$ where the I_k are mutually disjoint open intervals, is

$$L(O) = L(I_1) + L(I_2) + \cdots = \sum_{k=1}^{\infty} L(I_k) \tag{1}$$

If O is restricted to lie in some fundamental interval $I = [a, b]$, then the series in (1) converges to a non-negative number less than or equal to $b - a$, i.e.

$$0 \leqq L(O) \leqq b - a \tag{2}$$

29

LENGTH OF A CLOSED SET

Since a closed set C contained in $[a, b]$ is the complement of an open set, we are led to the following

Definition 2.4.   The length of a closed set $C \subset [a, b]$ is

$$L(C) = b - a - L(\tilde{C}) \tag{3}$$

THE CONCEPT OF MEASURE

A generalization of the concept of length to arbitrary sets on the real line leads to the concept of *measure of a set*. By analogy with length it would seem desirable to have the measure of a set E, denoted by $m(E)$, satisfy the following ideal requirements.

A-1. $m(E)$ is defined for each set E.

A-2. $m(E) \geqq 0$.

A-3   [**Finite additivity**].   If $E = \overset{n}{\underset{k=1}{\cup}} E_k$ where the E_k are mutually disjoint, then

$$m(E) = \sum_{k=1}^{n} m(E_k)$$

A-4   [**Denumerable additivity**].   If $E = \overset{\infty}{\underset{k=1}{\cup}} E_k$ where the E_k are mutually disjoint, then

$$m(E) = \sum_{k=1}^{\infty} m(E_k)$$

A-5   [**Monotonicity**].   If $E_1 \subset E_2$, then $m(E_1) \leqq m(E_2)$.

A-6   [**Translation invariance**].   If each point of a set E is translated equal distances in the same direction on the real line, the measure of the translated set is the same as $m(E)$.

A-7.   If E is an interval then $m(E) = L(E)$, the length of the interval.

Note that if requirement A-4 holds, then A-3 also holds. However, the converse need not be true. The two requirements A-3 and A-4 are sometimes together called the requirement of *countable additivity*. We can show that requirement A-5 follows from the other requirements [see Problem 2.31].

It is possible to show that for general point sets on the real line we cannot satisfy all of these requirements. Thus if we wish to keep all the requirements A-2 through A-7 we must conclude that not all sets have a measure. Similarly if we want all sets to have a measure, then we must sacrifice one or more of the requirements A-2 through A-7.

EXTERIOR OR OUTER MEASURE OF A SET

The *exterior* or *outer measure* of a set E, written $m_e(E)$, has the following properties which can be considered as axioms.

B-1. $m_e(E)$ is defined for each set E.

B-2. $m_e(E) \geqq 0$.

B-3. $m_e(E_1 \cup E_2 \cup \cdots) \leqq m_e(E_1) + m_e(E_2) + \cdots$

         for all sets $E_1, E_2, \ldots$ disjoint or not.

B-4.   Exterior measure is translation invariant.

Note that the requirements A-3 and A-4 for ideal measure, i.e. the additivity requirements, have been dropped and replaced by B-3.

MEASURABLE SETS

The exterior measure, although defined for all sets, does not satisfy in general the additivity requirements A-3 and A-4, as we have noted. Now we know that every set T can be written in the form [see Problem 1.10, page 12]

$$T = (T \cap E) \cup (T \cap \widetilde{E}) \qquad (4)$$

Thus if there is to be any chance at all of satisfying the additivity requirements, we ought to have

$$m_e(T) = m_e(T \cap E) + m_e(T \cap \widetilde{E}) \qquad (5)$$

for all sets T, since $T \cap E$ and $T \cap \widetilde{E}$ are disjoint. The result (5) will hold [if at all] only for a restrictive class of sets E. This leads to the following

Definition 2.5. A set E is said to be *measurable*, or more precisely *measurable with respect to an exterior measure*, if for all sets T

$$m_e(T) = m_e(T \cap E) + m_e(T \cap \widetilde{E}) \qquad (6)$$

In such case $m_e(E) = m(E)$ is called the *measure* of E.

We can show that if we restrict ourselves to measurable sets, then the above ideal requirements for measure are satisfied.

Since it is true that for all T [see Problem 2.8, page 36]

$$m_e(T) \leqq m_e(T \cap E) + m_e(T \cap \widetilde{E})$$

the Definition 2.5 can be replaced by the following equivalent one.

Definition 2.6. A set E is *measurable* if for all sets T

$$m_e(T) \geqq m_e(T \cap E) + m_e(T \cap \widetilde{E}) \qquad (7)$$

This is often used in practice to show that a set is measurable.

The sets T in the above definitions are often called *test sets* since they are used to test measurability.

LEBESGUE EXTERIOR OR OUTER MEASURE

Thus far we have not actually demonstrated the existence of an exterior or outer measure although there are in fact several which satisfy the axioms given above. The one which is most famous and useful is attributed to Lebesgue who investigated it in the early part of the 20th century.

Definition 2.7. The *Lebesgue exterior* or *outer measure* of a set E is

$$m_e(E) = \text{g.l.b. } L(O) \quad \text{for all open sets } O \supset E$$

i.e. the greatest lower bound of the lengths of all open sets O which contain E.

We can show that this does indeed satisfy the axioms for exterior measure [see Problem 2.4].

We can prove that if E is any interval I, then $m_e(I) = L(I)$. Also if S is any open set, then $m_e(S) = L(S)$. See Problems 2.2, 2.26 and 2.27.

The following theorem is important.

Theorem 2-1. If E is any given set, then given $\epsilon > 0$ there exists an open set $O \supset E$ such that $m_e(E) < m_e(O) + \epsilon$ and $m_e(O) \geqq m_e(E)$.

LEBESGUE MEASURE

Definition 2.8.   If a set E is measurable with respect to the Lebesgue exterior measure of Definition 2.7, then we say that E is *Lebesgue measurable* and define the *Lebesgue measure* of E as $m(E) = m_e(E)$.

In this book we shall, unless otherwise specified, be concerned only with *Lebesgue exterior measure* or *Lebesgue measure* of sets.   Consequently we shall often refer only to exterior measure or measure without mentioning the name Lebesgue explicitly.

In his investigations, Lebesgue did not actually use Definition 2.5 given above to define measurable sets.   Instead he considered sets E in the bounded interval $[a, b]$ and first defined the *interior* or *inner measure* of a set E as

$$m_i(E) = b - a - m_e(\widetilde{E}) \tag{8}$$

He then called E measurable if the interior and exterior measures are equal, i.e. if

$$m_e(E) = b - a - m_e(\widetilde{E}) \tag{9}$$

Now if we let $I = [a, b]$, (9) can be written

$$m_e(I) = m_e(I \cap E) + m_e(I \cap \widetilde{E}) \tag{10}$$

which is a special case of (6) with $T = I$.   Thus the actual definition which Lebesque used is a special case of (6).   Since Lebesgue started with sets contained in $[a, b]$, i.e. bounded sets, appropriate modifications had to be made for unbounded sets.   Such modifications, however, are not needed if Definition 2.5 is used.   Furthermore Definition 2.5 has the advantage that it can be used in more general theories of measure.   Consequently we shall use Definition 2.5 in various proofs.   It is of interest that if (10) is satisfied then (6) will also be satisfied [see Problem 2.18].

It should be mentioned that a modification of the Lebesgue procedure is sometimes used. According to this the interior measure of a set E is defined as

$$m_i(E) = \text{l.u.b.}\ L(C) \quad \text{for all closed sets } C \subset E \tag{11}$$

i.e. the least upper bound of the lengths of all closed sets C contained in E.   Then E is called measurable if $m_i(E) = m_e(E)$ where $m_e(E)$ is given by Definition 2.7.   Other procedures are also possible [see Problems 2.42-2.44].   All of these procedures can in fact be shown equivalent.

It should also be noted that for any set E, $m_i(E)$ and $m_e(E)$ always exist and

$$m_i(E) \leqq m_e(E) \tag{12}$$

This result is equivalent to (7) with $T = I$.

THEOREMS ON MEASURE

Theorem 2-2.   If E is measurable, then $\widetilde{E}$ is measurable and conversely.

Theorem 2-3.   If E has exterior measure zero, then E is measurable and $m(E) = 0$.

Theorem 2-4.   Any countable set has measure zero.

Theorem 2-5.   If E_1 and E_2 are measurable, then their union $E_1 \cup E_2$ is measurable.   The result can be extended to any countable union of sets [see Problems 2.10 and 2.12].

Theorem 2-6.   If E_1 and E_2 are measurable disjoint sets, then

$$m(E_1 \cup E_2) = m(E_1) + m(E_2)$$

This result can be generalized [see Problem 2.13].

Theorem 2-7. If E_1 and E_2 are measurable, then their intersection $E_1 \cap E_2$ is measurable. The result can be extended to any finite intersection of sets.

Theorem 2-8. If E_1 and E_2 are measurable and $E_1 \subset E_2$, then $m(E_1) \leqq m(E_2)$.

Theorem 2-9. If E_1 and E_2 are measurable, $E_1 \subset E_2$, and E_2 has finite measure, then $E_2 - E_1$ is measurable and $m(E_2 - E_1) = m(E_2) - m(E_1)$.

Theorem 2-10. If E_1 and E_2 are any measurable sets, then
$$m(E_1 \cup E_2) = m(E_1) + m(E_2) - m(E_1 \cap E_2)$$
This reduces to Theorem 2-6 if E_1 and E_2 are disjoint. The result can be generalized [see Problem 2.41].

Theorem 2-11. If $E_1, E_2, \ldots$ are measurable, then $E = \overset{\infty}{\underset{k=1}{\cup}} E_k$ is measurable.

Theorem 2-12. If $E_1, E_2, \ldots$ are mutually disjoint measurable sets, then
$$m(E_1 \cup E_2 \cup \cdots) = m(E_1) + m(E_2) + \cdots$$

Theorem 2-13. If $E_1, E_2, E_3, \ldots$ are measurable sets such that $E_1 \subset E_2 \subset E_3 \subset \cdots$, then $E = \overset{\infty}{\underset{k=1}{\cap}} E_k$ is measurable.

Theorem 2-14. If $E_1, E_2, E_3, \ldots$ are measurable sets such that $E_1 \subset E_2 \subset E_3 \subset \cdots$, then $E = \overset{\infty}{\underset{k=1}{\cup}} E_k$ is measurable and
$$m\left(\overset{\infty}{\underset{k=1}{\cup}} E_k \right) = \lim_{n \to \infty} m(E_n)$$

Theorem 2-15. If $E_1, E_2, E_3, \ldots$ are measurable sets such that $E_1 \supset E_2 \supset E_3 \supset \cdots$ and at least one of the E_k has finite measure, then $E = \overset{\infty}{\underset{k=1}{\cap}} E_k$ is measurable and
$$m\left(\overset{\infty}{\underset{k=1}{\cap}} E_k \right) = \lim_{n \to \infty} m(E_n)$$

Theorem 2-16. If I is any interval, $m(I) = L(I)$.

Theorem 2-17. If O is any open set, $m(O) = L(O)$.

Theorem 2-18. If C is any closed set, $m(C) = L(C)$.

ALMOST EVERYWHERE

A property which is true except for a set of measure zero is said to hold *almost everywhere*.

BOREL SETS

The class of sets which can be obtained by taking countable unions or intersections of open or closed sets is called the class of *Borel sets*. We have the following

Theorem 2-19. Any Borel set is measurable.

VITALI'S COVERING THEOREM

The following theorem due to *Vitali* is an important and interesting one which is reminiscent of the Heine-Borel theorem [see page 7].

Theorem 2-20 [Vitali].   Suppose that each point of a bounded, measurable set E is covered by a class $\mathcal{J}$ of intervals having arbitrarily small length [called a *Vitali covering*].   Then there exists a denumerable set of disjoint intervals $I_1, I_2, \ldots$ such that $\bigcup\limits_{k=1}^{\infty} I_k$ covers E except for a set of measure zero.

NON-MEASURABLE SETS

It would seem from the above results that all sets in R are measurable.   This however is not true, i.e. there are sets in R which are not measurable.   For an example of a non-measurable set see Problem 2.21.

Solved Problems

LENGTHS OF SETS

2.1.   Find the length of the set $\bigcup\limits_{k=1}^{\infty} \left\{ x : \dfrac{1}{k+1} \leqq x < \dfrac{1}{k} \right\}$.

Let $I_k = \left\{ x : \dfrac{1}{k+1} \leqq x < \dfrac{1}{k} \right\}$.   Then

$$I_1 = \{x : \tfrac{1}{2} \leqq x < 1\}, \quad I_2 = \{x : \tfrac{1}{3} \leqq x < \tfrac{1}{2}\}, \quad \ldots$$

so that $L(I_1) = 1 - \tfrac{1}{2}, \quad L(I_2) = \tfrac{1}{2} - \tfrac{1}{3}, \quad \ldots, \quad L(I_n) = \dfrac{1}{n} - \dfrac{1}{n+1}$

Then since the I_k are mutually disjoint,

$$L\left[\bigcup_{k=1}^{n} I_k \right] = \sum_{k=1}^{n} L(I_k) = (1 - \tfrac{1}{2}) + (\tfrac{1}{2} - \tfrac{1}{3}) + \cdots + \left(\dfrac{1}{n} - \dfrac{1}{n+1} \right) = 1 - \dfrac{1}{n+1}$$

Thus $L\left[\bigcup\limits_{k=1}^{\infty} I_k \right] = \lim\limits_{n \to \infty} L\left[\bigcup\limits_{k=1}^{n} I_k \right] = \lim\limits_{n \to \infty} \left(1 - \dfrac{1}{n+1} \right) = 1$

EXTERIOR OR OUTER MEASURE

2.2.   Prove that if S is any open set, then $m_e(S) = L(S)$.

Suppose that O is any open set containing S, i.e. $O \supset S$. Then each of the component intervals of S is contained in some component interval of O. Now S is the smallest open set which contains itself, i.e. $S \supset S$. Then by definition of exterior measure we have

$$m_e(S) = \text{g.l.b. } L(O) \quad \text{where } O \supset S = L(S)$$

2.3.   Prove Theorem 2-1, page 31: If E is any given set, then given $\epsilon > 0$ there exists an open set $O \supset E$ such that   (*a*) $m_e(E) < m_e(O) + \epsilon$ and   (*b*) $m_e(O) \geqq m_e(E)$.

(*a*) By definition of Lebesgue exterior measure,

$$m_e(E) = \text{g.l.b. } L(O) \quad \text{for all open sets } O \supset E$$

Since the greatest lower bound exists, it follows that $m_e(E)$ exists and by definition of greatest lower bound we will have for any $\epsilon > 0$

$$m_e(E) < L(O) + \epsilon$$

However, since $L(O) = m_e(O)$ by Problem 2.2, the required result follows.

(*b*) This follows from Problem 2.25.

2.4. Prove that the Lebesgue definition of exterior measure satisfies axiom B-3, page 30, for exterior measure.

We must show that for any sets $E_1, E_2, \ldots$

$$m_e\left(\bigcup_k E_k\right) \;\leq\; \sum_k m_e(E_k)$$

The inequality is trivial in case one of the sets has infinite exterior measure, so that we can restrict ourselves to the case where all sets have finite exterior measure.

In such case given $\epsilon > 0$ there are open sets $O_k \supset E_k$ where $k = 1, 2, \ldots$ such that [see Problem 2.3]

$$L(O_k) \;=\; m_e(O_k) \;<\; m_e(E_k) + \epsilon/2^k$$

$$L(O_k) \;=\; m_e(O_k) \;\geq\; m_e(E_k)$$

Then

$$m_e\left(\bigcup_k E_k\right) \;\leq\; m_e\left(\bigcup_k O_k\right) \;=\; \sum_k m_e(O_k) \;\leq\; \sum_k m_e(E_k) + \sum_k \epsilon/2^k \;\leq\; \sum_k m_e(E_k) + \epsilon$$

Thus since ϵ can be taken arbitrarily small,

$$m_e\left(\bigcup_k E_k\right) \;\leq\; \sum_k m_e(E_k)$$

We can also show that the other axioms on page 30 are satisfied.

2.5. Prove that if E is any countable set of real numbers, then $m_e(E) = 0$.

Let the points (real numbers) of the set be $a_1, a_2, a_3, \ldots$. Then we can enclose $a_1, a_2, a_3, \ldots$ in open intervals of lengths less than or equal to $\epsilon/2, \epsilon/2^2, \epsilon/2^3, \ldots$ respectively. Thus

$$m_e(E) \;\leq\; \frac{\epsilon}{2} + \frac{\epsilon}{2^2} + \frac{\epsilon}{2^3} + \cdots \;=\; \epsilon$$

Since ϵ can be taken arbitrarily small, it follows that $m_e(E) = 0$.

2.6. Prove that for each set T and any finite collection of mutually disjoint sets $E_1, \ldots, E_n$

$$m_e\left(T \cap \left\{\bigcup_{k=1}^{n} E_k\right\}\right) \;=\; \sum_{k=1}^{n} m_e(T \cap E_k)$$

We use mathematical induction. For $n = 1$ the result is certainly true. Assuming it to be true for $n - 1$ sets, we would have

$$m_e\left(T \cap \left\{\bigcup_{k=1}^{n-1} E_k\right\}\right) \;=\; \sum_{k=1}^{n-1} m_e(T \cap E_k) \tag{1}$$

Adding $m_e(T \cap E_n)$ to both sides yields

$$m_e(T \cap E_n) + m_e\left(T \cap \left\{\bigcup_{k=1}^{n-1} E_k\right\}\right) \;=\; \sum_{k=1}^{n} m_e(T \cap E_k) \tag{2}$$

This can be written [see Problem 2.30]

$$m_e\left(T \cap \left\{\bigcup_{k=1}^{n} E_k\right\} \cap E_n\right) + m_e\left(T \cap \left\{\bigcup_{k=1}^{n} E_k\right\} \cap \widetilde{E}_n\right) \;=\; \sum_{k=1}^{n} m_e(T \cap E_k) \tag{3}$$

Now since E_n is measurable, we have for each set T

$$m_e(T) \;=\; m_e(T \cap E_n) + m_e(T \cap \widetilde{E}_n) \tag{4}$$

Then letting $T = T \cap \left\{\bigcup_{k=1}^{n} E_k\right\}$, (3) becomes

$$m_e\left(T \cap \left\{\bigcup_{k=1}^{n} E_k\right\}\right) \;=\; \sum_{k=1}^{n} m_e(T \cap E_k) \tag{5}$$

Hence the result is true for n sets if it is true for $n - 1$ sets. But since it is true for 1 set it follows that it is true for 2 sets, 3 sets and so on.

MEASURABLE SETS AND THEOREMS ON MEASURE

2.7. Prove that if E is measurable, then the complement $\widetilde{E}$ is also measurable.

Since E is measurable, we have for each set T

$$m_e(T) = m_e(T \cap E) + m_e(T \cap \widetilde{E}) \tag{1}$$

In order that $\widetilde{E}$ be measurable we must show that for each set T

$$m_e(T) = m_e(T \cap \widetilde{E}) + m_e\{T \cap (\widetilde{E})^{\sim}\} = m_e(T \cap \widetilde{E}) + m_e(T \cap E) \tag{2}$$

But (2) is identical with (1).   Thus $\widetilde{E}$ is measurable.

2.8. Prove that Definitions 2.5 and 2.6, page 31, are equivalent.

For any sets T and E we have
$$T = (T \cap E) \cup (T \cap \widetilde{E}) \tag{1}$$

Thus by Problem 2.4, $m_e(T) \leqq m_e(T \cap E) + m_e(T \cap \widetilde{E}) \tag{2}$

Now according to Definition 2.5 a set E is measurable if

$$m_e(T) = m_e(T \cap E) + m_e(T \cap \widetilde{E}) \tag{3}$$

Thus we can conclude that a set E is measurable if for any set T

$$m_e(T) \geqq m_e(T \cap E) + m_e(T \cap \widetilde{E}) \tag{4}$$

which together with (2) yields (3).

Conversely if we adopt Definition 2.6, then in view of (2) we see that E is measurable according to Definition 2.5.

2.9. Prove that E is measurable if $m_e(E) = 0$.

For any set T we have $T \cap E \subset E$, so that

$$m_e(T \cap E) \leqq m_e(E) = 0$$

Then since $m_e(T \cap E) \geqq 0$, we have $m_e(T \cap E) = 0 \tag{1}$

For any set T we have $T \supset T \cap \widetilde{E}$, so that using (1),

$$m_e(T) \geqq m_e(T \cap \widetilde{E}) = m_e(T \cap E) + m_e(T \cap \widetilde{E}) \tag{2}$$

from which we see that E is measurable and $m(E) = m_e(E) = 0$.

2.10. Prove that if E_1 and E_2 are measurable, then $E_1 \cup E_2$ is measurable.

We would like to show that for any set T

$$m_e(T) \geqq m_e(T \cap [E_1 \cup E_2]) + m_e(T \cap [E_1 \cup E_2]^{\sim})$$

Since E_1 and E_2 are measurable, we have for any set T

$$\begin{aligned}
m_e(T) &= m_e(T \cap E_1) + m_e(T \cap \widetilde{E}_1) \\
&= m_e(T \cap E_1) + m_e[(T \cap \widetilde{E}_1) \cap E_2] + m_e[(T \cap \widetilde{E}_1) \cap \widetilde{E}_2] \\
&= m_e(T \cap E_1) + m_e(T \cap E_2 \cap \widetilde{E}_1) + m_e(T \cap \widetilde{E}_1 \cap \widetilde{E}_2) \\
&= m_e(T \cap E_1) + m_e(T \cap E_2 \cap \widetilde{E}_1) + m_e(T \cap [E_1 \cup E_2]^{\sim}) \\
&\geqq m_e(T \cap [E_1 \cup E_2]) + m_e(T \cap [E_1 \cup E_2]^{\sim})
\end{aligned}$$

Thus $E_1 \cup E_2$ is measurable.

2.11. Prove that if E_1 and E_2 are measurable, then $E_1 \cap E_2$ is measurable.

By Theorem 1-12, page 4, we have $E_1 \cap E_2 = (\widetilde{E}_1 \cup \widetilde{E}_2)^\sim$. Since E_1 and E_2 are measurable, it follows from Problems 2.7 and 2.10 that $E_1 \cap E_2$ is measurable.

2.12. Prove that a countable union of measurable sets is measurable.

We can always choose the sets in a countable union to be mutually disjoint [see Problem 1.35, page 18]. We must thus show that if $E_1, E_2, \ldots$ are mutually disjoint measurable sets, then $E = \overset{\infty}{\underset{k=1}{\cup}} E_k$ is measurable.

Now since any finite union of measurable sets is measurable, we have for any set T

$$m_e(T) = m_e\left(T \cap \left\{\overset{n}{\underset{k=1}{\cup}} E_k\right\}\right) + m_e\left(T \cap \left\{\overset{n}{\underset{k=1}{\cup}} E_k\right\}^\sim\right)$$

$$\geqq m_e\left(T \cap \left\{\overset{n}{\underset{k=1}{\cup}} E_k\right\}\right) + m_e\left(T \cap \left\{\overset{\infty}{\underset{k=1}{\cup}} E_k\right\}^\sim\right)$$

$$\geqq \overset{n}{\underset{k=1}{\sum}} m_e(T \cap E_k) + m_e(T \cap \widetilde{E})$$

Since $m_e(T)$ is independent of n, we have on taking the limit as $n \to \infty$

$$m_e(T) \geqq \overset{\infty}{\underset{k=1}{\sum}} m_e(T \cap E_k) + m_e(T \cap \widetilde{E}) \geqq m_e(T \cap E) + m_e(T \cap \widetilde{E})$$

Thus $E = \overset{\infty}{\underset{k=1}{\cup}} E_k$ is measurable.

2.13. Prove that if $E_1, E_2, \ldots$ are mutually disjoint measurable sets, then

$$m\left(\overset{\infty}{\underset{k=1}{\cup}} E_k\right) = \overset{\infty}{\underset{k=1}{\sum}} m(E_k)$$

This follows at once from Problem 2.12 by letting $T = R$, the set of real numbers.

2.14. Prove that the measure of any countable set E is zero.

By Problem 2.5, $m_e(E) = 0$. Thus by Problem 2.9, the measure of E is given by $m(E) = m_e(E) = 0$.

2.15. Prove that a countable union of countable sets has measure zero.

Since a countable union of countable sets is a countable set, the measure of this set is zero by Problem 2.14.

2.16. Given the measurable sets $E_1 \supset E_2 \supset E_3 \supset \cdots$ where $m(E_1)$ is finite. Prove that

$$m\left(\overset{\infty}{\underset{k=1}{\cap}} E_k\right) = \lim_{n \to \infty} m(E_n)$$

Let $E = \overset{\infty}{\underset{k=1}{\cap}} E_k$. Then we have

$$E_1 - E = (E_1 - E_2) \cup (E_2 - E_3) \cup \cdots \tag{1}$$

Since $E_1 - E_2, E_2 - E_3, \ldots$ are mutually disjoint and all the sets are measurable, we have by Problem 2.13

$$m(E_1 - E) = m(E_1 - E_2) + m(E_2 - E_3) + \cdots \tag{2}$$

Then since $E_1 \supset E$, $E_1 \supset E_2$, $E_2 \supset E_3$, $\ldots$, (2) can be written

$$
\begin{aligned}
m(E_1) - m(E) &= \lim_{n \to \infty} \left[m(E_1 - E_2) + m(E_2 - E_3) + \cdots + m(E_{n-1} - E_n) \right] \\
&= \lim_{n \to \infty} \left[m(E_1) - m(E_2) + m(E_2) - m(E_3) + \cdots + m(E_{n-1}) - m(E_n) \right] \\
&= \lim_{n \to \infty} \left[m(E_1) - m(E_n) \right] = m(E_1) - \lim_{n \to \infty} m(E_n)
\end{aligned}
$$

Thus since $m(E_1)$ is finite,

$$
m(E) = m\left(\bigcap_{k=1}^{\infty} E_k \right) = \lim_{n \to \infty} m(E_n)
$$

2.17. Prove that the Cantor set is measurable and find its measure.

In obtaining the Cantor set, the interval $[0, 1]$ is subdivided into 3 parts and the middle third is removed. Then the remaining intervals are in turn divided into 3 parts and the middle thirds of each are removed. The process is repeated indefinitely and what is left is the Cantor set.

Now at the first step 1 interval of length 1/3 is removed. At the second step 2 intervals of lengths $1/3^2$ are removed. In general after n steps, 2^{n-1} intervals of lengths $1/3^n$ are removed. It follows that the total measure of the Cantor set K is

$$
m(K) = 1 - \sum_{n=1}^{\infty} \frac{2^{n-1}}{3^n} = 1 - \frac{1/3}{1 - 2/3} = 0
$$

One of the interesting properties of the Cantor set is that although it is non-countable [see Problem 1.22, page 15] it has measure zero.

2.18. Let I be the interval $[a, b]$. Prove that a necessary and sufficient condition for a set E in I to be measurable is that

$$
m_e(I) = m_e(E) + m_e(\widetilde{E})
$$

Necessity. By definition a set E is measurable if for all sets T

$$
m_e(T) = m_e(T \cap E) + m_e(T \cap \widetilde{E})
$$

Then if in particular we let $T = I$, the required result follows since in this case $T \cap E = E$ and $T \cap \widetilde{E} = \widetilde{E}$.

Sufficiency. Let T be any set and suppose that G is a measurable set containing T such that $m_e(T) = m(G)$ [see Problem 2.45]. Since G is measurable, we have

$$
m_e(E) = m_e(E \cap G) + m_e(E \cap \widetilde{G})
$$
$$
m_e(\widetilde{E}) = m_e(\widetilde{E} \cap G) + m_e(\widetilde{E} \cap \widetilde{G})
$$

It follows that

$$
\begin{aligned}
m(I) &= m_e(E) + m_e(\widetilde{E}) \\
&= m_e(E \cap G) + m_e(E \cap \widetilde{G}) + m_e(\widetilde{E} \cap G) + m_e(\widetilde{E} \cap \widetilde{G}) \\
&= \left[m_e(E \cap G) + m_e(\widetilde{E} \cap G) \right] + \left[m_e(E \cap \widetilde{G}) + m_e(\widetilde{E} \cap \widetilde{G}) \right] \\
&\geqq m_e(G) + m_e(\widetilde{G}) = m(G) + m(\widetilde{G}) = m(I)
\end{aligned}
$$

We thus see that

$$
m_e(E \cap G) + m_e(E \cap \widetilde{G}) + m_e(\widetilde{E} \cap G) + m_e(\widetilde{E} \cap \widetilde{G}) = m(G) + m(\widetilde{G}) \tag{1}
$$

Now we know that

$$
m_e(E \cap \widetilde{G}) + m_e(\widetilde{E} \cap \widetilde{G}) \geqq m_e(\widetilde{G}) = m(G) \tag{2}
$$

Subtracting (2) from (1), $m_e(E \cap G) + m_e(\widetilde{E} \cap \widetilde{G}) \leqq m(G)$

Now since $T \subset G$, we have $E \cap T \subset E \cap G$ and $\widetilde{E} \cap T \subset \widetilde{E} \cap G$, so that

$$
m_e(E \cap T) + m_e(\widetilde{E} \cap T) \leqq m_e(E \cap G) + m_e(\widetilde{E} \cap G) \leqq m(G) = m_e(T)
$$

Then by Problem 2.8 it follows that E is measurable.

It should be noted that the theorem proved here is proved under the assumption that for any set T there is a measurable set G such that $m_e(T) = m(G)$. This is true for Lebesgue exterior measure as defined in Definition 2.7, page 31 [see Problem 2.45]. An exterior measure for which this is true is often called a *regular* exterior measure.

VITALI'S COVERING THEOREM

2.19. Prove Theorem 2-20 [Vitali's covering theorem].

Let O be an open set of finite measure containing E. Then we can assume that each interval of $\mathcal{J}$ belongs to O. Suppose that h_1 is the least upper bound of the lengths of all these intervals of $\mathcal{J}$. Choose an interval $I_1 \in \mathcal{J}$ whose length is greater than $\frac{1}{2}h_1$ i.e. $L(I_1) > \frac{1}{2}h_1$.

Now if $m_e(E - I_1) = 0$, the required result is established. If not, let h_2 be the least upper bound of the lengths of all those intervals of $\mathcal{J}$ which are disjoint from I_1. Denote by I_2 an interval of $\mathcal{J}$ which is disjoint from I_1 and whose length is greater than $\frac{1}{2}h_2$, i.e. $L(I_2) > \frac{1}{2}h_2$.

If $m_e(E - I_1 \cup I_2) = 0$, the required result follows. If not we continue the process to find an interval I_3, etc. Clearly the process will either terminate, in which case the result is established, or it will not terminate.

We must thus investigate the case where the process does not terminate. In such case we will have a sequence $I_1, I_2, \ldots$ of mutually disjoint intervals with the property that

$$m(I_k) = L(I_k) > \tfrac{1}{2}h_k, \qquad k = 1, 2, \ldots \tag{1}$$

Then since each $I_k \subset O$ so that $\bigcup_{k=1}^{\infty} I_k \subset O$, we have by Problem 2.13

$$m\left(\bigcup_{k=1}^{\infty} I_k\right) = \sum_{k=1}^{\infty} m(I_k) = \sum_{k=1}^{\infty} L(I_k) \leq m(O) \tag{2}$$

Thus the series in (2) converges and as a consequence of this and (1),

$$\lim_{k \to \infty} L(I_k) = 0 \quad \text{and} \quad \lim_{k \to \infty} h_k = 0$$

Consider now the set $S = E - \bigcup_{k=1}^{n} I_k$ which is not empty. If $x \in S$, then it belongs to E and does not belong to $I_1, I_2, \ldots, I_n$. By the definition of Vitali covering there will be some interval $I \in \mathcal{J}$ containing x. Also I is disjoint from $I_1, \ldots, I_n$.

The interval I must have at least one point in common with some interval I_k where $k > n$. For if I were disjoint from $I_{n+1}, I_{n+2}, \ldots$ for $k = n+1, n+2, \ldots$, it would follow on noting that $L(I) \leq h_k$ for $k = n+1, n+2, \ldots$ and $\lim_{k \to \infty} h_k = 0$ that $L(I) = 0$ so that I would not be an interval.

We can assume that the first interval in the sequence $\langle I_k \rangle$ with which I has common points is I_{n+1}. We see that in such case

$$L(I) \leq h_{n+1} > 2L(I_{n+1}) \tag{1}$$

Since $x \in I$, we see from Fig. 2-1 that the distance from x to the midpoint of I_{n+1} cannot exceed $L(I) + \frac{1}{2}L(I_{n+1})$ or $\frac{5}{2}L(I_{n+1})$ using (1). If therefore we consider an interval J_{n+1} with the same midpoint as I_{n+1} but five times as long, we see that $x \in J_{n+1}$.

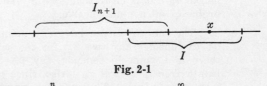

Fig. 2-1

From this it follows that all points x belonging to $E - \bigcup_{k=1}^{n} I_k$ also belong to $\bigcup_{k=n+1}^{\infty} J_k$, i.e.

$$E - \bigcup_{k=1}^{n} I_k \subset \bigcup_{k=n+1}^{\infty} J_k \tag{2}$$

Then by Problem 2.25,

$$m_e\left(E - \bigcup_{k=1}^{n} I_k\right) \leq \sum_{k=n+1}^{\infty} L(J_k) = 5\sum_{k=n+1}^{\infty} L(I_k)$$

But since $\sum\limits_{k=1}^{\infty} L(I_k)$ converges, it follows that given $\epsilon > 0$ there is a number n_0 such that the last sum can be made less than $\epsilon/5$ for $n > n_0$, i.e.

$$m_e\left(E - \bigcup_{k=1}^{n} I_k\right) < \epsilon \quad \text{for} \quad n > n_0 \tag{3}$$

2.20. Prove that the conclusion of Problem 2.19 is the same as the statement

$$m\left(E - \bigcup_{k=1}^{\infty} I_k\right) = 0$$

or that the mutually disjoint intervals $I_1, I_2, \ldots$ cover E almost everywhere.

The result (3) of Problem 2.19 can be written equivalently as

$$m_e\left(E - \bigcup_{k=1}^{\infty} I_k\right) = 0$$

Thus from Problem 2.9 we have

$$m\left(E - \bigcup_{k=1}^{\infty} I_k\right) = 0$$

NON-MEASURABLE SETS

2.21. Demonstrate the existence of a non-measurable set.

We shall show that there is a non-measurable set in the interval $[0, 1]$. For convenience in arriving at this set we shall map this interval on to the circumference C of a circle so that to each point in $[0, 1]$ there will be one and only one point on C and conversely.

Let x and y denote any two points on C. We define x and y to be equivalent points if the arc joining them has rational length and in such case write $x \sim y$.

We now consider subsets of C, denoted by A_α, having the property that two points x and y are in the same A_α if $x \sim y$. The sets A_α are often called *equivalence classes*. Since the rational numbers are countable, it is then clear that each A_α contains a countably infinite set of points, i.e. is countable.

The sets A_α are also mutually disjoint. For if the same point x belongs to different sets A_{α_1} and A_{α_2}, then all points of A_{α_1} and A_{α_2} are the same so that A_{α_1} is identical with A_{α_2}.

Since the sets A_α are mutually disjoint, it follows that there is a non-countable number of such sets. To see this we simply have to observe that each set must contain a distinct irrational number and that the set of irrational numbers is non-countable.

We now construct a set S which consists of one and only one point x_α from each of the sets A_α so that any two distinct points of S are not equivalent. To do this we must use the axiom of choice. The set S is the required non-measurable set.

To show that S is non-measurable let us first consider the set of rational numbers between 0 and 1, denoted by $r_1, r_2, r_3, \ldots$. For simplicity let $r_1 = 0$. For any positive integer k let S_k be the set of points of C which we obtain by a counterclockwise rotation of S through an arc of length r_k. It is clear that $S = S_1$ and all the S_k are congruent, so that they are either all measurable or all non-measurable.

The sets S_k do not have any points in common, i.e. they are mutually disjoint. To see this let us consider any two S_k, for the sake of argument S_4 and S_7. If S_4 and S_7 are not disjoint there is a point in common, say p. Since $p \in S_4$ there is an element $y \in S$ for which the arc length from p to y is r_4. Also, since $p \in S_7$ there is an element $z \in S$ for which the arc length from p to z is r_7. Thus the arc length from y to z is rational, i.e. $y \sim z$. This however contradicts the fact that no two distinct points of S can be equivalent, and thus we must have $y = z$ so that we must have $r_4 = r_7$ which is impossible. This contradiction shows that the S_k are mutually disjoint.

Now if we take any point $x \in C$, then we have $x \in A_\alpha$ for some α. Since $x_\alpha \in A_\alpha$, it follows that the arc length from x to x_α equals r_k for some positive integer k so that $x \in S_k$. Thus each point of C belongs to some S_k.

We thus conclude that $C = \bigcup\limits_{k=1}^{\infty} S_k$ where the S_k are mutually disjoint.   Then if the S_k are measurable we have $m(S) = m(S_1) = m(S_2) = \cdots$ and

$$m(C) \;=\; m(S_1) + m(S_2) + \cdots \;=\; m(S) + m(S) + \cdots$$

Now if $m(S) > 0$ we would have $m(C) = \infty$, and if $m(S) = 0$ we would have $m(C) = 0$. Since $m(C) = 1$, we cannot have $m(C) = \infty$ or 0. This contradiction shows that S cannot be measurable as we were required to show.

Supplementary Problems

LENGTHS OF SETS

2.22.   Find the lengths of each of the following sets.

  (a) $\{x : -2 < x < 6\}$ (b) $\{x : 3 \leqq x \leqq 5\} \cup \{-4 \leqq x < -2\}$ (c) $\{x : -3 < x < 4\} \cup \{1 \leqq x \leqq 6\}$

  Ans. (a) 8   (b) 4   (c) 9

2.23.   Find the length of the set $\bigcup\limits_{k=1}^{\infty} \left\{ x : \dfrac{1}{2^k} \leqq x < \dfrac{1}{2^{k-1}} \right\}$.   *Ans.* 1

2.24.   Find the length of the set $\bigcup\limits_{k=1}^{\infty} \left\{ x : 0 < x < \dfrac{1}{3^k} \right\}$.   *Ans.* 1/3

EXTERIOR OR OUTER MEASURE

2.25.   If $E_1 \subset E_2$, prove that $m_e(E_1) \leqq m_e(E_2)$.

2.26.   Prove that if I is a closed interval $[a, b]$, then $m_e(I) = L(I) = b - a$.

2.27.   Extend the result of Problem 2.26 to any interval.

2.28.   Prove that if C is any closed set, then $m_e(C) = L(C)$.

2.29.   If $m_e(E_1) = 0$, prove that $m_e(E_1 \cup E_2) = m_e(E_2)$.

2.30.   Verify equation (3) of Problem 2.6.

MEASURABLE SETS AND THEOREMS ON MEASURE

2.31.   Prove that requirement A-5 for measure follows from the other requirements.

2.32.   Prove that any subset of the rational numbers Q is measurable and find its measure.

2.33.   Prove that the set of all irrational numbers in $[0, 1]$ is measurable and find its measure.   Is any subset of the irrational numbers measurable?   Justify your answer.

2.34.   Prove that every interval I is measurable and has a measure equal to its length, i.e. $m(I) = L(I)$.

2.35.   Prove that if O is any open set, then O is measurable and $m(O) = L(O)$.

2.36.   Prove that if C is any closed set, then C is measurable and $m(C) = L(C)$.

2.37.   Prove Theorem 2-8, page 33.

2.38.   Prove Theorem 2-10, page 33.

2.39.   Prove or disprove: Any subset of a measurable set is measurable.

2.40.   Use Problem 2.5 to prove that the set of points in $[0,1]$ is not countable.

2.41.   If E_1, E_2, E_3 are any measurable sets, prove that
$$m(E_1 \cup E_2 \cup E_3) \; = \; m(E_1) + m(E_2) + m(E_3) - m(E_1 \cap E_2) - m(E_1 \cap E_3) - m(E_2 \cap E_3) + m(E_1 \cap E_2 \cap E_3)$$
which is a generalization of Theorem 2-10, page 33.   Obtain further generalizations.

2.42.   Prove that a set E is measurable if and only if for any $\epsilon > 0$ there exists an open set $O \supset E$ such that $m_e(O - E) < \epsilon$.

2.43.   Prove that a set E is measurable if and only if for any $\epsilon > 0$ there exists a closed set $C \subset E$ such that $m_e(E - C) < \epsilon$.

2.44.   Prove that a set E is measurable if and only if for any $\epsilon > 0$ there exist an open set $O \supset E$ and a closed set $C \subset E$ such that $m_e(O - C) < \epsilon$.

2.45.   If A is any set, prove that there is a measurable set $B \supset A$ such that $m_e(A) = m(B)$.   See Problem 2.18.

2.46.   Suppose that all the numbers between 0 and 1 are expressed as non-terminating expansions in the scale of 10, i.e. as infinite decimals.   Prove that the measure of the set of all such numbers in which one particular digit [say 5] is omitted is zero.

2.47.   Prove Theorem 2-19, page 33.

2.48.   Use Lebesgue's definition of a measurable set [see (8) or (9), page 32] to prove (a) Theorem 2-6, (b) Theorem 2-11.

2.49.   Show how to prove some of the theorems on pages 32 and 33 if we start with the definition that E is measurable if there is an open set $O \supset E$ such that $m_e(O - E) < \epsilon$ [compare Problem 2.42].

2.50.   Work Problem 2.49 if we use the definition that E is measurable if there is an open set $O \supset E$ and a closed set $C \subset E$ such that $m_e(O - C) < \epsilon$ [compare Problem 2.44].

Chapter 3

Measurable Functions

DEFINITION OF A MEASURABLE FUNCTION

Let E be a measurable set and $f(x)$ a real function defined on E. We say that $f(x)$ is *Lebesgue measurable* or, briefly, *measurable* on E if for each real number κ the set of values $x \in E$ for which $f(x) > \kappa$ is measurable. The set of values $x \in E$ for which $f(x) > \kappa$ can be written $\{x \in E : f(x) > \kappa\}$ or briefly $E[f(x) > \kappa]$.

If $f(x)$ is measurable we call it a *measurable function*.

THEOREMS ON MEASURABLE FUNCTIONS

Theorem 3-1. The function $f(x)$ is measurable on E if and only if for each real number κ one of the following sets is measurable

$$(a)\ E[f(x) < \kappa], \quad (b)\ E[f(x) \leqq \kappa], \quad (c)\ E[f(x) \geqq \kappa]$$

and if one of these sets is measurable then all the sets are measurable.

Because of this theorem a function $f(x)$ can be defined as measurable on E if any one of the sets $E[f(x) > \kappa]$, $E[f(x) < \kappa]$, $E[f(x) \leqq \kappa]$, $E[f(x) \geqq \kappa]$ is measurable for each κ.

Theorem 3-2. If $f(x)$ is measurable on E, then $E[f(x) = \kappa]$ is measurable for each κ. The converse of this is not true [see Problem 3.6].

Theorem 3-3. The function $f(x)$ is measurable on E if and only if for each pair of distinct real numbers α and β, $E[\alpha < f(x) < \beta]$ is measurable. The result is also true if either one or both of the inequality symbols is replaced by $\leqq$.

Theorem 3-4. A function $f(x)$ is measurable on E if $E[f(x) > \kappa]$ is measurable for each *rational number* κ.

Theorem 3-5. If $f(x)$ is measurable on a set E_1 and if $E_2 \subset E_1$, then $f(x)$ is measurable on E_2.

Theorem 3-6. If $f(x)$ is measurable on a countable class of disjoint sets $E_1, E_2, \ldots$, then it is measurable on their union $\bigcup_{k=1}^{\infty} E_k$.

Theorem 3-7. If $f_1(x)$ and $f_2(x)$ are measurable on E, then $E[f_1(x) > f_2(x)]$ is measurable.

Theorem 3-8. A constant function is measurable.

Theorem 3-9. If $f(x)$ is measurable on E, then for any constant c, $f(x) + c$ and $c f(x)$ are also measurable on E.

Theorem 3-10. If $f(x)$ is measurable on E, then $[f(x)]^2$ is measurable on E.

Theorem 3-11. If $f_1(x)$ and $f_2(x)$ are measurable on E, then $f_1(x) + f_2(x)$, $f_1(x) - f_2(x)$, $f_1(x) f_2(x)$ and $f_1(x)/f_2(x)$ where $f_2(x) \neq 0$ are measurable on E.

Theorem 3-12. If $f_1(x)$ and $f_2(x)$ are measurable on E, then the maximum and minimum of $f_1(x)$ and $f_2(x)$, i.e. max $\{f_1(x), f_2(x)\}$ and min $\{f_1(x), f_2(x)\}$, are measurable.

Theorem 3-13. If $f(x)$ is measurable on E, then $|f(x)|$ is measurable on E.

Theorem 3-14. If $f(x)$ is continuous on E, it is measurable on E.

Theorem 3-15. If $f_1(u)$ is continuous and $u = f_2(x)$ is measurable, then $f_1(f_2(x))$ is measurable. In other words a continuous function of a measurable function is measurable. However, a measurable function of a measurable function need not be measurable.

Theorem 3-16. Let $\langle f_n(x) \rangle$ be a sequence of functions measurable on E. Then $F(x) =$ l.u.b. $f_n(x)$, called the *upper boundary function*, and $G(x) =$ g.l.b. $f_n(x)$, called the *lower boundary function*, are also measurable on E.

Theorem 3-17. If $\langle f_n(x) \rangle$ is a monotonic sequence of functions measurable on E such that $\lim_{n \to \infty} f_n(x) = f(x)$, then $f(x)$ is measurable on E. [Note: The sequence $\langle f_n(x) \rangle$ is said to be *monotonic increasing* if $f_1(x) \leq f_2(x) \leq \cdots$, *monotonic decreasing* if $f_1(x) \geq f_2(x) \geq \cdots$, and *monotonic* if the sequence is monotonic increasing *or* monotonic decreasing.]

Theorem 3-18. Let $\langle f_n(x) \rangle$ be a sequence of functions measurable on E. Then $\overline{\lim}_{n \to \infty} f_n(x)$ and $\underline{\lim}_{n \to \infty} f_n(x)$ are measurable on E.

Theorem 3-19. Let $\langle f_n(x) \rangle$ be a sequence of functions measurable on E such that $\lim_{n \to \infty} f_n(x) = f(x)$. Then $f(x)$ is measurable on E. The result is also valid in case $\lim_{n \to \infty} f_n(x) = f(x)$ almost everywhere.

Theorem 3-20. If $f_1(x)$ is measurable on a set E and $f_1(x) = f_2(x)$ almost everywhere on E, then $f_2(x)$ is measurable on E.

BAIRE CLASSES

From the above theorems we see that all continuous functions are measurable and all limits of sequences of measurable functions are measurable. It follows that limits of sequences of continuous functions, which may be discontinuous, are measurable and that limits of these discontinuous functions are measurable, etc. Thus we obtain a hierarchy of measurable functions.

This has led *Baire* to give a classification of functions. A function is said to belong to the *Baire class of order zero* [or briefly *Baire class 0*] if it is continuous. A function which is the limit of a sequence of continuous functions [i.e. functions belonging to Baire class 0] but which is itself not continuous, belongs to the *Baire class of order one* [or briefly *Baire class 1*]. In general a function is said to belong to *Baire class p* if it is the limit of a sequence of functions of Baire class $p - 1$ but does not itself belong to any of the Baire classes $0, 1, \ldots, p - 1$.

It follows that every function which belongs to some Baire class is measurable.

EGOROV'S THEOREM

Theorem 3-21 [Egorov].   Let $\langle f_n(x) \rangle$ be a sequence of measurable functions which converges to a finite limit $f(x)$ almost everywhere on a set E of finite measure.   Then given any number $\delta > 0$, there exists a set F of measure greater than $m(E) - \delta$ on which $f_n(x)$ converges to $f(x)$ uniformly.

Solved Problems

3.1.   Prove that $f(x)$ is measurable on E if and only if for each real number κ the set $E[f(x) \leqq \kappa]$ is measurable.

   If $f(x)$ is measurable on E, then for each real number κ the set $E[f(x) > \kappa]$ is measurable. Then by Theorem 2-2, page 32, its complement with respect to E given by $E[f(x) \leqq \kappa]$ is measurable.

   Conversely if $E[f(x) \leqq \kappa]$ is measurable, so also is $E[f(x) > \kappa]$, i.e. $f(x)$ is measurable on E.

3.2.   Prove that $f(x)$ is measurable on E if and only if for each real number κ the set $E[f(x) \geqq \kappa]$ is measurable.

   If $f(x)$ is measurable on E, then for each real number κ the set $E[f(x) > \kappa]$ is measurable.   Then $E\left[f(x) > \kappa - \dfrac{1}{n} \right]$ is measurable for $n = 1, 2, 3, \ldots$ and so

$$\bigcap_{n=1}^{\infty} E\left[f(x) > \kappa - \frac{1}{n} \right] \; = \; E[f(x) \geqq \kappa]$$

is measurable.

   Conversely if $E[f(x) \geqq \kappa]$ is measurable, then $E\left[f(x) \geqq \kappa + \dfrac{1}{n} \right]$ is measurable for $n = 1, 2, 3, \ldots$ and so

$$\bigcup_{n=1}^{\infty} E\left[f(x) \geqq \kappa + \frac{1}{n} \right] \; = \; E[f(x) > \kappa]$$

is measurable, i.e. $f(x)$ is measurable on E.

3.3.   Prove that $f(x)$ is measurable on E if and only if for each real number κ the set $E[f(x) < \kappa]$ is measurable.

   If $f(x)$ is measurable on E, then $E[f(x) > \kappa]$ is measurable for each real number κ.   Then by Problem 3.2, $E[f(x) \geqq \kappa]$ is measurable and so its complement with respect to E given by $E[f(x) < \kappa]$ is measurable.

   Conversely if $E[f(x) < \kappa]$ is measurable, then its complement with respect to E given by $E[f(x) \geqq \kappa]$ is measurable and so by Problem 3.2 $f(x)$ is measurable on E.

3.4.   Prove that if $f(x)$ is measurable on E, then $E[f(x) = \kappa]$ is measurable for each real number κ.

   We have
$$E[f(x) = \kappa] \; = \; E[f(x) \geqq \kappa] \cap E[f(x) \leqq \kappa]$$

Since the sets on the right are measurable, their intersection is also measurable and so the result follows.

3.5. Prove that if $f(x)$ is measurable on E, then   (a) $E[f(x) = \infty]$ and   (b) $E[f(x) = -\infty]$ are measurable.

(a) We have
$$E[f(x) = \infty] = \bigcap_{\kappa=1}^{\infty} E[f(x) > \kappa]$$

and since a countable intersection of measurable sets is measurable the required result follows.

(b) We have
$$E[f(x) = -\infty] = \bigcap_{\kappa=1}^{\infty} E[f(x) < -\kappa]$$

and the required result follows as in part (a).

3.6. Give an example to show that even if $E[f(x) = \kappa]$ is measurable for each real number κ, it may be true that $f(x)$ is not measurable on E.   Compare Problem 3.4.

Suppose that S is a set which is non-measurable [see Problem 2.21, page 40, for example] and let this set be made to correspond in a 1-1 manner with some subset of the set of points for which $x > 0$.   Suppose further that the complement of S, i.e. $\widetilde{S}$, corresponds in a 1-1 manner with some subset of the set of points for which $x \leqq 0$.

Now let us suppose that a function $f(x)$ is defined by these 1-1 correspondences.   Then if κ is any real number, the set $E[f(x) = \kappa]$ contains not more than one point and is thus measurable. However, $E[f(x) > 0]$ which is the same as S is non-measurable, i.e. $f(x)$ is not measurable.

3.7. Prove Theorem 3-4, page 43:   A function $f(x)$ is measurable on E if   (a) $E[f(x) > \kappa]$ is measurable for each rational number κ or   (b) $E[f(x) \geqq \kappa]$ is measurable for each rational number κ.

(a) Let κ be the limit of a sequence of rational numbers $\langle \kappa_n \rangle$ where $\kappa_1 > \kappa_2 > \kappa_3 > \cdots$.   We have
$$E[f(x) > \kappa] = \bigcup_{n=1}^{\infty} E[f(x) > \kappa_n]$$

Then since each of the sets on the right is measurable, their countable union is also measurable and so $f(x)$ is measurable on E.

(b) This follows as in part (a), since we can write
$$E[f(x) \geqq \kappa] = \bigcup_{n=1}^{\infty} E[f(x) \geqq \kappa_n]$$

3.8. Prove Theorem 3-7, page 43: If $f_1(x)$ and $f_2(x)$ are measurable on E, then $E[f_1(x) > f_2(x)]$ is measurable.

For every $x \in E$ we can always find a rational number r such that $f_1(x) > r > f_2(x)$.   Since
$$E[f_1(x) > r > f_2(x)] = E[f_1(x) > r] \cap E[f_2(x) < r] \tag{1}$$

it follows that the set on the left is measurable since the sets on the right are measurable.

Now we have
$$E[f_1(x) > f_2(x)] = \bigcup_{r} E[f_1(x) > r > f_2(x)] \tag{2}$$

where the union on the right is taken over all rational numbers r such that $f_1(x) > r > f_2(x)$. Since the set of rational numbers is countable, the right side of (2) is a countable union of measurable sets which is measurable.

3.9. Prove that if $f(x)$ is measurable on E, then for any real constant c, $f(x) + c$ is measurable on E.

Since $f(x)$ is measurable on E, we have $E[f(x) > \kappa - c]$ is measurable.   Thus $E[f(x) + c > \kappa]$ is measurable and so $f(x) + c$ is measurable on E.

3.10. Prove that if $f(x)$ is measurable on E, then for any real constant c, $c\,f(x)$ is measurable on E.

If $c > 0$, $E[c\,f(x) > \kappa] = E[f(x) > \kappa/c]$ which is measurable.

If $c < 0$, $E[c\,f(x) > \kappa] = E[f(x) < \kappa/c]$ which is measurable.

If $c = 0$, $c\,f(x) = 0$ which is measurable.

Thus $c\,f(x)$ is measurable on E for any real constant c.

3.11. Prove that if $f_1(x)$ and $f_2(x)$ are measurable on E, then $f_1(x) + f_2(x)$ is measurable on E.

Method 1.

We have for each real number κ,

$$E[f_1(x) + f_2(x) > \kappa] = E[f_1(x) > \kappa - f_2(x)] \tag{1}$$

By Problem 3.9 and 3.10, $\kappa - f_2(x)$ is measurable. Thus by Problem 3.8 the right side of (1) is measurable, so that $f_1(x) + f_2(x)$ is measurable on E.

Method 2.

We have for each real number κ,

$$E[f_1(x) + f_2(x) > \kappa] = \bigcup_r E[f_1(x) > r] \cap E[f_2(x) > \kappa - r] \tag{2}$$

where the union on the right is taken over all rational numbers r and is thus a countable union. Since $f_1(x)$ and $f_2(x)$ are measurable on E, so also are the sets on the right of (2) and a countable union of these sets. Thus the left side of (2) is measurable and so $f_1(x) + f_2(x)$ is measurable on E.

3.12. Prove Theorem 3-10, page 44: If $f(x)$ is measurable on E, then $\{f(x)\}^2$ is measurable on E.

We have

$$E[\{f(x)\}^2 > \kappa] = E[f(x) > \sqrt{\kappa}] \cup E[f(x) < -\sqrt{\kappa}]$$

and the required result follows since the union of measurable sets is measurable.

3.13. Prove that if $f_1(x)$ and $f_2(x)$ are measurable on E, then $f_1(x)\,f_2(x)$ is measurable on E.

We have

$$f_1(x)\,f_2(x) = \tfrac{1}{4}[\{f_1(x) + f_2(x)\}^2 - \{f_1(x) - f_2(x)\}^2]$$

Thus by Problems 3.10 and 3.11, $f_1(x)\,f_2(x)$ is measurable on E.

3.14. Prove Theorem 3-14, page 44: If $f(x)$ is continuous on E, it is measurable on E.

Consider the set $E[f(x) \leq \kappa]$. If x_0 is any limit point of this set, then every neighborhood of x_0 contains points such that $f(x) \leq \kappa$. By the continuity of $f(x)$ at x_0, it follows that $f(x_0) \leq \kappa$. We thus see that the limit point x_0 belongs to the set $E[f(x) \leq \kappa]$. Then this set is closed since it contains all its limit points. It follows that $E[f(x) \leq \kappa]$ is measurable since any closed set is measurable. Thus the complement $E[f(x) > \kappa]$ is measurable and so $f(x)$ is measurable on E.

3.15. Let $f(x)$ be measurable on E and let O be an open set. Prove that the set $E[f(x) \in O]$ is measurable.

Let $O = \bigcup_{k=1}^{\infty} I_k$ where $I_k = (a_k, b_k)$ are the component open disjoint intervals [see Theorem 1-19, page 7]. It follows that

$$E[f(x) \in O] \;=\; \bigcup_{k=1}^{\infty} E[f(x) \in I_k] \;=\; \bigcup_{k=1}^{\infty} \left(E[f(x) > a_k] \cap E[f(x) < b_k] \right)$$

Since the sets $E[f(x) > a_k]$ and $E[f(x) < b_k]$ are measurable, it follows that $E[f(x) \in O]$ is measurable.

3.16. Prove Theorem 3-15, page 44: If $f_1(u)$ is continuous and $u = f_2(x)$ is measurable on E, prove that $f_1(f_2(x))$ is measurable on E.

We must show that $E[f_1(f_2(x)) > \kappa]$ is measurable. Now

$$E[f_1(f_2(x)) > \kappa] \;=\; E[f_2(x) \in O] \tag{1}$$

where

$$O \;=\; \{ u : f_1(u) > \kappa \}$$

is open since f_1 is continuous [see Problem 1.41]. Then since the right side of (1) is measurable, the required result follows.

3.17. Prove Theorem 3-16, page 44: Let $\langle f_n(x) \rangle$ be a sequence of functions measurable on E. Then $F(x) = \text{l.u.b.} \{ f_n(x) \} = \text{l.u.b.} \{ f_1(x), f_2(x), \ldots \}$ and $G(x) = \text{g.l.b.} \{ f_n(x) \} = \text{g.l.b.} \{ f_1(x), f_2(x), \ldots \}$ are also measurable on E.

We have

$$E[F(x) \geqq \kappa] \;=\; \bigcup_{n=1}^{\infty} E[f_n(x) \geqq \kappa]$$

Then since a countable union of measurable sets is also measurable, the required result follows.

A similar argument shows that $G(x) = \text{g.l.b.}\, f_n(x)$ is measurable on E [see Problem 3.45].

3.18. If $f_1(x)$ and $f_2(x)$ are measurable on E, prove that (a) $\max \{ f_1(x), f_2(x) \}$ and (b) $\min \{ f_1(x), f_2(x) \}$, where max and min denote maximum and minimum, are also measurable on E.

(a) We have

$$\max \{ f_1(x), f_2(x) \} \;=\; \text{l.u.b.} \{ f_1(x), f_2(x) \}$$

so that the required result follows from Problem 3.17.

(b) We have

$$\min \{ f_1(x), f_2(x) \} \;=\; \text{g.l.b.} \{ f_1(x), f_2(x) \}$$

so that the required result follows from Problem 3.17.

3.19. Prove Theorem 3-17, page 44: If $\langle f_n(x) \rangle$ is a monotonic sequence of functions measurable on E such that $\lim\limits_{n \to \infty} f_n(x) = f(x)$, then $f(x)$ is measurable on E.

Consider the case where for all $x \in E$,

$$f_1(x) \;\geqq\; f_2(x) \;\geqq\; f_3(x) \;\geqq\; \cdots$$

We have for each real number κ

$$E\left[\lim_{n \to \infty} f_n(x) \geqq \kappa \right] \;=\; \bigcup_{k=1}^{\infty} E[f_k(x) > \kappa]$$

and the required result follows since a countable union of measurable sets is measurable.

A similar proof can be given for the case where $f_1(x) \leqq f_2(x) \leqq f_3(x) \leqq \cdots$ or cases where the equality is omitted.

3.20. Prove Theorem 3-18, page 44: Let $\langle f_n(x)\rangle$ be a sequence of functions measurable on E. Then $\varlimsup_{n\to\infty} f_n(x)$ and $\varliminf_{n\to\infty} f_n(x)$ are measurable on E.

Call
$$F_1(x) = \text{l.u.b. } \{f_1(x), f_2(x), \ldots\}$$
$$F_2(x) = \text{l.u.b. } \{f_2(x), f_3(x), \ldots\}$$
$$F_3(x) = \text{l.u.b. } \{f_3(x), f_4(x), \ldots\}$$

and so on. It follows that $F_1(x) \geqq F_2(x) \geqq F_3(x) \geqq \cdots$ and
$$\varlimsup_{n\to\infty} f_n(x) = \lim_{n\to\infty} F_n(x)$$

Now by Problem 3.19, the right side is measurable and so the required result follows.

A similar argument shows that $\varliminf_{n\to\infty} f_n(x)$ is measurable [see Problem 3.46].

3.21. Prove Theorem 3-19, page 44: Let $\langle f_n(x)\rangle$ be a sequence of functions measurable on E such that $\lim_{n\to\infty} f_n(x) = f(x)$. Then $f(x)$ is measurable on E.

By Problem 3.20, $\varlimsup_{n\to\infty} f_n(x)$ and $\varliminf_{n\to\infty} f_n(x)$ are measurable on E. If $\lim_{n\to\infty} f_n(x) = f(x)$, then
$$f(x) = \varlimsup_{n\to\infty} f_n(x) = \varliminf_{n\to\infty} f_n(x)$$
and so is measurable on E.

3.22. Prove that there are functions which are non-measurable.

Let S be a non-measurable set [see Problem 2.21] and consider the *characteristic function*
$$\chi(x) = \begin{cases} 1 & \text{if } x \in S \\ 0 & \text{if } x \notin S \end{cases}$$
Then $\chi(x)$ is non-measurable.

BAIRE CLASSES

3.23. Let $f_n(x) = \dfrac{1}{1 + x^{2n}}$ for any real number x. (a) To what Baire class does each $f_n(x)$ belong? (b) To what Baire class does $\lim_{n\to\infty} f_n(x)$ belong?

(a) Since each $f_n(x)$ is continuous, it belongs to the Baire class of order zero, i.e. Baire class 0.

(b) For $|x| < 1$, $\lim_{n\to\infty} f_n(x) = 1$. For $|x| > 1$, $\lim_{n\to\infty} f_n(x) = 0$. For $|x| = 1$, i.e. $x^{2n} = 1$, $\lim_{n\to\infty} f_n(x) = \frac{1}{2}$. Thus if $\lim_{n\to\infty} f_n(x) = F(x)$, then
$$F(x) = \begin{cases} 1 & |x| < 1 \\ \frac{1}{2} & |x| = 1 \\ 0 & |x| > 1 \end{cases}$$

Since $F(x)$ is the limit of a sequence of functions in Baire class zero but does not itself belong to Baire class zero, it belongs to Baire class 1.

3.24. Prove that the sum of two functions of Baire class p is also of Baire class p.

The result is true if $p = 0$, since the sum of two continuous functions is also continuous.

To prove that the result is true for $p = 1$, we observe that by definition if $f(x)$ and $g(x)$ belong to Baire class 1 then there are sequences of functions $\langle f_n(x)\rangle$ and $\langle g_n(x)\rangle$ such that $\lim_{n\to\infty} f_n(x) = f(x)$

and $\lim\limits_{n \to \infty} g_n(x) = g(x)$ where each $f_n(x)$ and $g_n(x)$ belongs to Baire class 0. Now since $f_n(x) + g_n(x)$ belongs to Baire class 0 it follows that $f(x) + g(x) = \lim\limits_{n \to \infty} [f_n(x) + g_n(x)]$ belongs to Baire class 1.

The same argument can be used to prove the result for any value of p.

EGOROV'S THEOREM

3.25. Prove Theorem 3-21 [Egorov's theorem], page 45: Let $\langle f_n(x) \rangle$ be a sequence of measurable functions which converges to a finite limit $f(x)$ almost everywhere on a set E of finite measure. Then given any number $\delta > 0$, there exists a set F of measure greater than $m(E) - \delta$ on which $f_n(x)$ converges to $f(x)$ uniformly.

Suppose that the set of all $x \in E$ for which $f_n(x)$ converges to $f(x)$ is denoted by H. Then we have

$$m(H) = m(E) \quad \text{or} \quad m(E - H) = 0 \tag{1}$$

Let

$$H_{p,j} = \left\{ x : x \in H, |f_n - f| < \frac{1}{p} \text{ for } n \geqq j \right\}$$

Now if we fix p it follows that

$$H_{p,1} \subset H_{p,2} \subset H_{p,3} \subset \cdots$$

and

$$H = \bigcup_{j=1}^{\infty} H_{p,j}$$

Then by Theorem 2-14, page 33, we have

$$\lim_{j \to \infty} m(H_{p,j}) = m(H)$$

so that there must be a natural number $j(p)$ for which

$$m(H - H_{p,j(p)}) < \frac{\delta}{2^p} \tag{2}$$

If we call

$$F = \bigcap_{p=1}^{\infty} H_{p,j(p)}$$

we see that on F

$$|f_n - f| < \frac{1}{p} \quad \text{for all } n \geqq j(p)$$

i.e. $f_n(x)$ converges to $f(x)$ uniformly on F.

We now have only to show that $m(F) > m(E) - \delta$ or equivalently $m(E - F) < \delta$. To do this note that

$$H - F \subset \bigcup_{p=1}^{\infty} (H - H_{p,j(p)})$$

so that by (1), (2) and Theorem 2-8, we have

$$m(E - F) = m(H - F) \leqq \sum_{p=1}^{\infty} m(H - H_{p,j(p)}) < \sum_{p=1}^{\infty} \frac{\delta}{2^p} = \delta$$

i.e. $m(E - F) < \delta$ as required.

Supplementary Problems

3.26. Prove Theorem 3-3, page 43.

3.27. Prove a theorem analogous to that of Theorem 3-3, page 43, where either one or both of the inequality symbols $<$ is replaced by $\leqq$.

3.28. Prove that a constant function is measurable.

3.29. Investigate the measurability of the function
$$f(x) \;=\; \begin{cases} 1 & \text{if } x \text{ is a rational number in } (0,1) \\ 0 & \text{if } x \text{ is an irrational number in } (0,1) \end{cases}$$

3.30. Prove that if $f_1(x)$ and $f_2(x)$ are measurable, then $c_1 f_1(x) + c_2 f_2(x)$, where c_1 and c_2 are any constants, is also measurable.

3.31. Prove that $f(x)$ is measurable on E if and only if $E[f(x) < \kappa]$ is measurable for each rational number κ.

3.32. Prove that any function defined on a set of measure zero is measurable.

3.33. Prove that the sum and product of any finite number of measurable functions is also measurable.

3.34. Is the result of Problem 3.33 true for an infinite number of measurable functions? Explain.

3.35. Prove Theorem 3-5, page 43.

3.36. Prove Theorem 3-6, page 43.

3.37. Prove that if $f_1(x)$ and $f_2(x)$ are measurable on E, then $f_1(x)/f_2(x)$ is measurable on E if $f_2(x) \neq 0$. [*Hint*: First prove that $1/f_2(x)$ is measurable].

3.38. If $f(x)$ is measurable, prove that any positive integral power of $f(x)$ is also measurable.

3.39. Prove Theorem 3-12, page 44.

3.40. Prove Theorem 3-13, page 44.

3.41. Prove Theorem 3-19, page 44, for the case where $\lim\limits_{n \to \infty} f_n(x) = f(x)$ almost everywhere in E.

3.42. Prove Theorem 3-20, page 44.

3.43. Prove that the function $f(x)$ defined in the interval $[a, b]$ by
$$f(x) \;=\; \begin{cases} c_1 & a \leqq x < q \\ c_2 & q \leqq x \leqq b \end{cases}$$
where c_1 and c_2 are constants, is measurable in $[a, b]$. This function is an example of a *step function*.

3.44. Suppose that $[a, b]$ is the union of a countable number of disjoint sets $E_1, E_2, \ldots$ so that $f(x) = c_k$ on E_k where c_k is a given constant. Prove that $f(x)$ is measurable in $[a, b]$. Note that if $E_1, E_2, \ldots$ are intervals, then $f(x)$ is a *step function* [Problem 3.43].

3.45. Complete Problem 3.17.

3.46. Complete Problem 3.20.

3.47. Let $f(x)$ be measurable and $g(x)$ be monotonic increasing [i.e. $g(x)$ is such that $g(x_2) \geqq g(x_1)$ for $x_2 > x_1$]. Prove that $g(f(x))$ is measurable.

3.48. Prove the result of Problem 3.47 if $g(x)$ is monotonic decreasing [i.e. $g(x_2) \leqq g(x_1)$ for $x_2 > x_1$]. Thus prove the result if $g(x)$ is monotonic [i.e. monotonic increasing *or* monotonic decreasing].

3.49. Show that $\lim\limits_{p \to \infty} \lim\limits_{q \to \infty} (\cos p! \, \pi x)^{2q}$ is measurable.

3.50. To what Baire class does the function of Problem 3.49 belong?

3.51. Let $f_n(x) = e^{-nx^2}$ for $0 \leqq x \leqq 1$ and let $F(x) = \lim\limits_{n \to \infty} f_n(x)$. To what Baire classes do the functions (a) $f_n(x)$ and (b) $F(x)$ belong?

3.52. Let S be a non-measurable set. Suppose that for each $p \in S$ we define a function $f_p(x) = \begin{cases} 1 & x = p \\ 0 & \text{otherwise} \end{cases}$. (a) Show that $F(x) = \text{l.u.b.} \, [f_p(x) \text{ where } p \in S] = \begin{cases} 1 & \text{if } x \in S \\ 0 & \text{otherwise} \end{cases}$ and that (b) $F(x)$ is non-measurable.

3.53. Does the result of Problem 3.52 contradict Theorem 3-16, page 44?

3.54. Prove the result of Problem 3.24 for the case where $p = 2$.

3.55. Prove that if $f(x)$ is of Baire class p, then so also are (a) $|f(x)|$, (b) $[f(x)]^2$.

3.56. Prove that the product of two functions of Baire class p is also of Baire class p.

3.57. Prove that if $f_1(x)$ and $f_2(x)$ are of Baire class p, then so also are (a) $\max\{f_1(x), f_2(x)\}$ and (b) $\min\{f_1(x), f_2(x)\}$. [*Hint*: $\max\{f_1(x), f_2(x)\} = \frac{1}{2}[f_1(x) + f_2(x)] + \frac{1}{2}|f_1(x) - f_2(x)|$, $\min\{f_1(x), f_2(x)\} = \frac{1}{2}[f_1(x) + f_2(x)] - \frac{1}{2}|f_1(x) - f_2(x)|$.]

3.58. Use Problem 3.57 to work Problem 3.18.

3.59. Let $\langle f_n(x) \rangle$ be a sequence of functions of Baire class p which converges uniformly to $f(x)$. Prove that $f(x)$ is also of Baire class p.

3.60. Is the result of Problem 3.59 true in case the convergence is not uniform? Explain.

Chapter 4

The Lebesgue Integral for Bounded Functions

THE RIEMANN INTEGRAL

The integral usually treated in introductory courses in the calculus is called the *Riemann integral* after the mathematician who contributed largely to its development. A discussion of this integral and its properties is given in Appendix A, pages 154-174, for the benefit of those who wish a review of it.

The Riemann integral has certain defects which can be remedied by use of the *Lebesgue integral,* as we shall see in this and later chapters. Although it is possible to define the Lebesgue integral in many ways, we shall adopt a procedure which parallels as closely as possible the definition given in Appendix A for the Riemann integral. The main difference between the Riemann and Lebesgue integrals is that the former uses intervals and their lengths while the latter uses more general point sets and their measures. Thus it is not surprising that the Lebesgue integral is more general than the Riemann integral.

DEFINITION OF LEBESGUE INTEGRAL FOR BOUNDED MEASURABLE FUNCTIONS

Let $f(x)$ be bounded and measurable on the interval $[a, b]$. Suppose that α and β are any two real numbers such that $\alpha < f(x) < \beta$. Divide the range α to β into n subintervals by choosing values $y_1, y_2, \ldots, y_{n-1}$ so that

$$\alpha = y_0 < y_1 < y_2 < \cdots < y_{n-1} < y_n = \beta \tag{1}$$

Note that these values are represented geometrically by points on the y axis as indicated in Fig. 4-1. A particular set of subdivision points is often referred to as a *partition, net* or *mode of subdivision.*

Let E_k, $k = 1, 2, \ldots, n$, be the set of all x in $[a, b]$ such that $y_{k-1} \leqq f(x) < y_k$, i.e.

$$E_k = \{x : y_{k-1} \leqq f(x) < y_k\}, \qquad k = 1, 2, \ldots, n$$

Since $f(x)$ is measurable, these sets are measurable and, as easily verified, disjoint.

Consider the *upper* and *lower sums* S and s respectively defined by

$$S = \sum_{k=1}^{n} y_k \, m(E_k) \tag{2}$$

$$s = \sum_{k=1}^{n} y_{k-1} \, m(E_k) \tag{3}$$

By varying the partition, i.e. choosing different points of subdivision as well as the number of points, we obtain sets of values for S and s. Let

$I = $ g.l.b. of the values of S for all possible partitions

$J = $ l.u.b. of the values of s for all possible partitions

These values, which always exist, are called *upper* and *lower Lebesgue integrals* of $f(x)$ on $[a, b]$ respectively and are denoted by

$$I = \overline{\int_a^b} f(x)\, dx, \qquad J = \underline{\int_a^b} f(x)\, dx \tag{4}$$

If $I = J$, we say that $f(x)$ is *Lebesgue integrable* on $[a, b]$ and denote the common value by

$$\int_a^b f(x)\, dx \tag{5}$$

called the *Lebesgue definite integral* of $f(x)$ on $[a, b]$.

We have the following important

Theorem 4-1.   If $f(x)$ is bounded and measurable on $[a, b]$, then the Lebesgue integral (5) exists and we say that $f(x)$ is *Lebesgue integrable* or simply *integrable* on $[a, b]$.

If (5) exists, we sometimes write

$$\int_a^b f(x)\, dx \, < \, \infty \tag{6}$$

GEOMETRIC INTERPRETATION OF THE LEBESGUE INTEGRAL

As seen in Fig. 4-1 below, $y_k\, m(E_k)$ and $y_{k-1}\, m(E_k)$ can be interpreted geometrically as areas of rectangles when the E_k are intervals on the x axis. In such case the Lebesgue integral gives the area bounded by the curve $y = f(x)$, the x axis and the ordinates at $x = a$ and $x = b$. For more general sets the Lebesgue integral may exist but a geometric interpretation may be impossible.

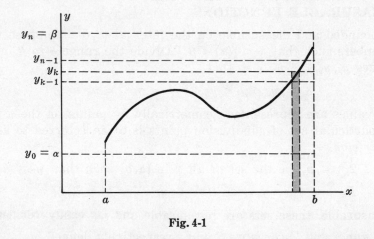

Fig. 4-1

NOTATION FOR RIEMANN AND LEBESGUE INTEGRALS

It should be observed that (5) has the same appearance as the Riemann integral of elementary calculus. Since the main purpose of this book is a treatment of the Lebesgue integral, we shall use the notation (5) for this integral. When it is necessary to distinguish between this integral and the Riemann integral, we shall use the notation

$$(\mathcal{R}) \int_a^b f(x)\, dx \tag{7}$$

for the Riemann integral. [An exception to this is Appendix A.]

THE LEBESGUE INTEGRAL DEFINED ON A BOUNDED MEASURABLE SET

If E is a measurable set contained in $[a, b]$, then the Lebesgue integral of $f(x)$ on E is defined as

$$\int_E f(x)\,dx = \int_a^b g(x)\,dx \tag{8}$$

where

$$g(x) = \begin{cases} f(x) & \text{for } x \in E \\ 0 & \text{for } x \notin E \end{cases} \tag{9}$$

This can also be defined directly through use of the upper and lower sums (2) and (3) where

$$E_k = \{x: x \in E, \; y_{k-1} \leqq f(x) < y_k\} \tag{10}$$

If this is done, then the results already obtained become a special case where $E = [a, b]$.

THE LEBESGUE INTEGRAL AS A LIMIT OF A SUM

We can also define the Lebesgue integral as the common limit of the sums S and s given by (2) and (3) respectively as the number of subdivision points becomes infinite in such a manner that the largest value of $y_k - y_{k-1}$ approaches zero. See Problems 4.11 and 4.39.

THEOREMS INVOLVING THE LEBESGUE INTEGRAL

In the following we assume, unless otherwise stated, that $f(x)$ is bounded and measurable and thus Lebesgue integrable and that all sets indicated are measurable.

Theorem 4-2. $\displaystyle\int_E c\,f(x)\,dx = c\int_E f(x)\,dx$ for any constant c

Theorem 4-3. $\displaystyle\int_E c\,dx = c\,m(E)$ for any constant c

Theorem 4-4. If E has measure zero, then

$$\int_E f(x)\,dx = 0$$

Theorem 4-5. If $A \leqq f(x) \leqq B$, then

$$Am(E) \leqq \int_E f(x)\,dx \leqq Bm(E) \tag{11}$$

This is sometimes called the *mean-value theorem* for Lebesgue integrals.

Theorem 4-6. If $E = E_1 \cup E_2$ where E_1 and E_2 are disjoint, then

$$\int_E f(x)\,dx = \int_{E_1} f(x)\,dx + \int_{E_2} f(x)\,dx$$

The result is easily generalized to finite unions of sets.

Theorem 4-7. If $E = E_1 \cup E_2 \cup \cdots$ where $E_1, E_2, \ldots$ are mutually disjoint, then

$$\int_E f(x)\,dx = \int_{E_1} f(x)\,dx + \int_{E_2} f(x)\,dx + \cdots$$

This generalizes Theorem 4-6 to a countably infinite union of sets.

Theorem 4-8.
$$\int_E [f(x) + g(x)]\, dx = \int_E f(x)\, dx + \int_E g(x)\, dx$$

The result is easily generalized to any finite number of functions.

Theorem 4-9. If $f(x)$ and $g(x)$ are bounded and measurable on E, then $f(x)\, g(x)$ is Lebesgue integrable on E, i.e.
$$\int_E f(x)\, g(x)\, dx < \infty$$

Theorem 4-10. If $f(x) \leq g(x)$ on E, then
$$\int_E f(x)\, dx \leq \int_E g(x)\, dx$$

The result is also true if $f(x) \leq g(x)$ almost everywhere on E.

Theorem 4.11. If $f(x)$ is bounded and measurable on E, then $|f(x)|$ is Lebesgue integrable on E. Conversely if $|f(x)|$ is bounded and measurable on E, then $f(x)$ is Lebesgue integrable on E.

Theorem 4-12.
$$\left| \int_E f(x)\, dx \right| \leq \int_E |f(x)|\, dx$$

under the conditions of Theorem 4-11.

Theorem 4-13. If $f(x) = g(x)$ almost everywhere on E, then
$$\int_E f(x)\, dx = \int_E g(x)\, dx$$

Theorem 4-14. If $f(x) \geq 0$ and $\int_E f(x)\, dx = 0$, then $f(x) = 0$ almost everywhere on E. The result is also true if $f(x) \geq 0$ almost everywhere on E.

LEBESGUE'S THEOREM ON BOUNDED CONVERGENCE

Theorem 4-15. [**Bounded convergence**]. Let $\langle f_n(x) \rangle$ be a sequence of functions measurable on E such that $\lim_{n \to \infty} f_n(x) = f(x)$. Then if the sequence is uniformly bounded, i.e. if there is a constant M such that $|f_n(x)| \leq M$ for all n, we have
$$\lim_{n \to \infty} \int_E f_n(x)\, dx = \int_E \lim_{n \to \infty} f_n(x)\, dx = \int_E f(x)\, dx$$

Because sets of measure zero can be omitted in the integrals, the conditions of the theorem can be relaxed so that they hold almost everywhere.

This important theorem, which is not true for Riemann integrals [see Problem 4.24], indicates the superiority of the Lebesgue integral.

THE BOUNDED CONVERGENCE THEOREM FOR INFINITE SERIES

The Lebesgue theorem on bounded convergence stated above can be restated in terms of series rather than sequences as follows.

Theorem 4-16. If $u_k(x), k = 1, 2, \ldots$ are measurable on E and the partial sums $s_n(x) = \sum_{k=1}^{n} u_k(x)$ are uniformly bounded on E $\left[\text{i.e. } \left| \sum_{k=1}^{n} u_k(x) \right| \leq M \text{ for all } x \in E \right]$ and $\lim_{n \to \infty} s_n(x) = s(x)$, then
$$\int_E \left\{ \sum_{k=1}^{\infty} u_k(x) \right\} dx = \sum_{k=1}^{\infty} \int_E u_k(x)\, dx$$

RELATIONSHIP OF RIEMANN AND LEBESGUE INTEGRALS

Theorem 4-17. If $f(x)$ is Riemann integrable in $[a, b]$, then it is also Lebesgue integrable in $[a, b]$ and the two integrals are equal, i.e.

$$\int_a^b f(x)\,dx \;=\; (\mathcal{R})\int_a^b f(x)\,dx$$

Note, however, that the converse is not true, i.e. if $f(x)$ is Lebesgue integrable in $[a, b]$ then it need not be Riemann integrable in $[a, b]$. See Problem 4.9 and also Problem A.2, page 159.

Theorem 4-18. A function $f(x)$ is Riemann integrable in $[a, b]$ if and only if the set of discontinuities of $f(x)$ in $[a, b]$ has measure zero, i.e. if $f(x)$ is continuous almost everywhere.

Theorem 4-19. If $f(x)$ is continuous almost everywhere in $[a, b]$, then it is Lebesgue integrable in $[a, b]$.

Solved Problems

DEFINITION OF THE LEBESGUE INTEGRAL FOR BOUNDED MEASURABLE FUNCTIONS

4.1. Prove that for the same mode of subdivision or partition, a lower sum s is not greater than an upper sum S, i.e. $s \leqq S$.

Using the notation on page 53, we have

$$y_{k-1} < y_k$$

Then since $m(E_k) \geqq 0$, 　　　　　　　　$y_{k-1}\,m(E_k) \;\leqq\; y_k\,m(E_k)$

Summing from $k = 1$ to n, we have

$$s \;=\; \sum_{k=1}^n y_{k-1}\,m(E_k) \;\leqq\; \sum_{k=1}^n y_k\,m(E_k) \;=\; S$$

4.2. Prove that for the same mode of subdivision or partition (a) an upper sum has a lower bound and (b) a lower sum has an upper bound.

(a) For $k = 1, \ldots, n$ we have $y_k \geqq \alpha$ so that

$$y_k\,m(E_k) \;\geqq\; \alpha\,m(E_k)$$

Summing from $k = 1$ to n yields

$$S \;=\; \sum_{k=1}^n y_k\,m(E_k) \;\geqq\; \alpha \sum_{k=1}^n m(E_k) \;=\; \alpha\,m(E)$$

where $E = \bigcup\limits_{k=1}^n E_k = \{x : \alpha < f(x) < \beta\}$.

Then a lower bound of S is $\alpha\,m(E)$.

(b) For $k = 1, \ldots, n$ we have $y_{k-1} \leqq \beta$ so that

$$y_{k-1}\, m(E_k) \;\leqq\; \beta\, m(E_k)$$

Summing from $k = 1$ to n yields

$$s \;=\; \sum_{k=1}^{n} y_{k-1}\, m(E_k) \;\leqq\; \beta \sum_{k=1}^{n} m(E_k) \;=\; \beta\, m(E)$$

Then an upper bound of s is $\beta\, m(E)$.

4.3. Prove that for all possible modes of subdivision or partitions (a) the upper sums have a greatest lower bound I and (b) the lower sums have a least upper bound J.

(a) By Problem 4.2(a), the upper sums for all possible modes of subdivision have a lower bound $\alpha\, m(E)$ and must thus have a greatest lower bound which we can denote by I.

(b) By Problem 2(b), the lower sums for all possible modes of subdivision have an upper bound $\beta\, m(E)$ and thus must have a least upper bound which we can denote by J.

4.4. A *refinement* of a given partition or mode of subdivision is obtained by using additional points of subdivision. If S is the upper sum corresponding to a given partition while S_1 is the upper sum corresponding to a refinement of this partition, prove that $S_1 \leqq S$. Thus prove that in a refinement of a partition, upper sums cannot increase. Similarly we can show that $s_1 \geqq s$, i.e. the lower sums cannot decrease [see Problem 4.32].

The result will be proved if we can prove it when one point of subdivision is added to the given partition. To do this let the given subdivision points be

$$\alpha \;=\; y_0 \;<\; y_1 \;<\; \cdots \;<\; y_n \;=\; \beta$$

Suppose that the additional point of subdivision occurs in the interval (y_{p-1}, y_p) and is denoted by u so that $y_{p-1} < u < y_p$.

Now the contribution to the upper sum corresponding to the subdivision points of the interval (y_{p-1}, y_p) is

$$y_p\, m(E_p) \tag{1}$$

The contribution to the upper sum when the additional point of subdivision u is taken into account is

$$u\, m(E_p^{(1)}) \;+\; y_p\, m(E_p^{(2)}) \tag{2}$$

where $$E_p^{(1)} \;=\; \{x : y_{p-1} \leqq f(x) < u\}, \qquad E_p^{(2)} \;=\; \{x : u \leqq f(x) < y_p\} \tag{3}$$

and it is clear that

$$E_p \;=\; E_p^{(1)} \cup E_p^{(2)}, \qquad m(E_p) \;=\; m(E_p^{(1)}) + m(E_p^{(2)}) \tag{4}$$

Because of (4) we can write (1) as

$$y_p\, m(E_p^{(1)}) \;+\; y_p\, m(E_p^{(2)}) \tag{5}$$

The change in the original upper sum caused by the additional subdivision point is given by the difference between (2) and (5), i.e.

$$(u - y_p)\, m(E_p^{(1)}) \tag{6}$$

Since this is negative or zero [because $y_p > u$ and $m(E_p^{(1)}) \geqq 0$], it follows that the upper sum cannot increase by adding a point of subdivision and the required result is proved.

4.5. Prove that an upper sum for any partition is not less than a lower sum corresponding to the same or any other partition.

The required result has already been proved when the partitions are the same [Problem 4.1].

To prove the result for different partitions, suppose that the upper and lower sums corresponding to one partition are given by S_2, s_2 while the upper and lower sums corresponding to the other partition are given by S_3, s_3.

Consider the partition obtained by superimposing the two above partitions [i.e. the partition obtained by taking the union of all the subdivision points] and denote the corresponding upper and lower sums by S_4, s_4. Since this new partition is a refinement of each of the given partitions, we have by Problem 4.4,

$$S_2 \geqq S_4, \qquad s_4 \geqq s_3 \tag{1}$$

But we also have by Problem 4.1,

$$S_4 \geqq s_4 \tag{2}$$

Thus from (1) and (2) we have

$$S_2 \geqq s_3 \tag{3}$$

[or $S_3 \geqq s_2$ on interchanging subscripts 2 and 3], i.e. an upper sum for any partition is not less than a lower sum for any other partition.

4.6. Using the notation of Problem 4.3, prove that $I \geqq J$.

Assume the contrary, i.e. $I < J$. Now I is the greatest lower bound of all upper sums. Thus there must be some partition or mode of subdivision giving an upper sum S to the left of the midpoint of the interval IJ [Fig. 4-2].

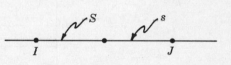

Also since J is the least upper bound of all lower sums, there must be some partition giving a sum s to the right of the midpoint of interval IJ.

Fig. 4-2

This, however, is impossible because S cannot be less than s. Thus we have arrived at a contradiction on assuming $I > J$, and it follows that $I \leqq J$.

Note that although we have used a geometric argument the proof can be easily formulated analytically [see Problem 4.33].

4.7. If S is an upper sum corresponding to any partition and s is a lower sum corresponding to the same or different partition, prove that

$$S \geqq I \geqq J \geqq s$$

Since I is the greatest lower bound of all upper sums, we must have $S \geqq I$. Since J is the least upper bound of all lower sums, we must have $J \geqq s$. Also from Problem 4.6 we have $I \geqq J$. Thus $S \geqq I \geqq J \geqq s$.

4.8. Prove that $f(x)$ is Lebesgue integrable if and only if for any $\epsilon > 0$ there exists a partition with upper and lower sums S, s such that $S - s < \epsilon$.

If for any $\epsilon > 0$ there is a partition such that $S - s < \epsilon$, then by Problem 4.7,

$$0 \leqq I - J \leqq S - s < \epsilon$$

so that $I = J$ [since ϵ can be arbitrarily small]. Then $f(x)$ is Lebesgue integrable.

Conversely suppose $I = J$. Then if $\epsilon > 0$ there exists a partition such that $S < I + \epsilon/2$ [since I is the greatest lower bound of all upper sums] and $s > J - \epsilon/2$ [since J is the least upper bound of all lower sums]. Then

$$S - s < (I + \epsilon/2) - (J - \epsilon/2) = \epsilon$$

4.9. (*a*) Show directly from the definition that the function

$$f(x) = \begin{cases} 1 & x \text{ is a rational number in } [0,1] \\ 0 & x \text{ is an irrational number in } [0,1] \end{cases}$$

is Lebesgue integrable in $[0,1]$ and (*b*) find the value of the Lebesgue integral of $f(x)$ in $[0,1]$.

(*a*) In the definition of the Lebesgue integral [see page 53] we choose subdivision points such that

$$\alpha = y_0 < y_1 < y_2 < \cdots < y_{n-1} < y_n = \beta$$

and consider sets $\qquad E_k = \{x : y_{k-1} \leqq f(x) < y_k\}$

Now let us consider Fig. 4-3, where the heavy lines are intended to indicate the graph of $f(x)$ in $[0,1]$, and the dashed lines are intended to show the subdivision points taken on the y axis [compare Fig. 4-1, page 54]. Although it is certainly not necessary to appeal to a diagram, the use of one helps to clarify the ideas involved.

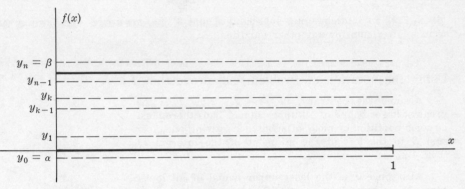

Fig. 4-3

It is clear from this diagram that the only non-empty sets are [assuming $\alpha \leqq 0,\ \beta > 1$, $y_1 > 0,\ y_{n-1} < 1$]

$$E_1 = \{x : y_0 \leqq f(x) < y_1\}$$

and $\qquad E_n = \{x : y_{n-1} \leqq f(x) < y_n\}$

The set E_1 is the set of irrational numbers in $[0,1]$ while the set E_n is the set of rational numbers in $[0,1]$. It follows therefore that $m(E_1) = 1$ and $m(E_n) = 0$ while $m(E_k) = 0$ for $k = 2, 3, \ldots, n-1$.

Then since an upper sum and lower sum are defined respectively by

$$S = \sum_{k=1}^{n} y_k\, m(E_k), \qquad s = \sum_{k=1}^{n} y_{k-1}\, m(E_k)$$

we see that these reduce to

$$S = y_1, \quad s = y_0$$

Now by varying the partition it is then clear that since y_1 is any number > 0 while $y_0 = \alpha$ is any number $\leqq 0$,

$$I = \text{l.u.b.}\,S = \text{l.u.b.}\,y_1 = 0$$

$$J = \text{g.l.b.}\,s = \text{g.l.b.}\,y_0 = 0$$

Thus since $I = J = 0$ we see that $f(x)$ is Lebesgue integrable in $[0,1]$.

(*b*) Since the common value of the upper and lower integrals I and J is 0, it follows that the Lebesgue integral of $f(x)$ in $[0,1]$ is 0, i.e.

$$\int_0^1 f(x)\, dx = 0$$

Note that the function is not Riemann integrable [see Problem A.2, page 159].

4.10. Prove that if $f(x)$ is bounded and measurable on $[a, b]$, then it is Lebesgue integrable on $[a, b]$.

Using the notation on page 53, we have

$$S - s = \sum_{k=1}^{n} (y_k - y_{k-1}) \, m(E_k)$$

Now we can certainly arrange to have a partition or mode of subdivision so that $y_k - y_{k-1} < \dfrac{\epsilon}{b-a}$. Thus we can say that

$$S - s < \sum_{k=1}^{n} \frac{\epsilon}{b-a} \, m(E_k) = \frac{\epsilon}{b-a} \sum_{k=1}^{n} m(E_k) = \epsilon$$

so that from Problem 4.8, $f(x)$ is Lebesgue integrable.

THE LEBESGUE INTEGRAL AS THE LIMIT OF A SUM

4.11. Prove that the definition of the Lebesgue integral as a limit of a sum [see page 55] follows from the definition given on page 53.

For the case where the number of subdivision points is n, let S_n and s_n be the corresponding upper and lower sums. Then if I is the g.l.b. of all upper sums and J is the l.u.b. of all lower sums, we must have

$$J_n \leqq J \leqq I \leqq I_n \qquad\qquad (1)$$

Now

$$S_n = \sum_{k=1}^{n} y_k \, m(E_k), \qquad s_n = \sum_{k=1}^{n} y_{k-1} \, m(E_k)$$

so that

$$0 \leqq S_n - s_n = \sum_{k=1}^{n} (y_k - y_{k-1}) \, m(E_k) \qquad\qquad (2)$$

Then given $\epsilon > 0$ we can choose $n > n_0$ so that $y_k - y_{k-1} < \epsilon/(b-a)$ for $k = 1, 2, \ldots, n$. Thus for $n > n_0$,

$$S_n - s_n \leqq \sum_{k=1}^{n} \frac{\epsilon}{b-a} m(E_k) \leqq \epsilon$$

i.e. $\displaystyle \lim_{n \to \infty} (S_n - s_n) = 0$. Thus from (1) we see that

$$\lim_{n \to \infty} I_n = \lim_{n \to \infty} J_n = I = J = \int_a^b f(x)\, dx$$

We can also show the converse [see Problem 4.39].

THEOREMS INVOLVING THE LEBESGUE INTEGRAL

4.12. Prove Theorem 4-5, page 55: If $A \leqq f(x) \leqq B$, then

$$A \, m(E) \leqq \int_E f(x)\, dx \leqq B \, m(E)$$

From Problems 4.2 and 4.7,

$$A \, m(E) \leqq s \leqq I \leqq J \leqq S \leqq B \, m(E)$$

Since we are assuming that $f(x)$ is Lebesgue integrable so that $I = J$, it follows that

$$A \, m(E) \leqq \int_E f(x)\, dx \leqq B \, m(E)$$

4.13. Prove Theorem 4-6, page 55: If $E = E_1 \cup E_2$ where E_1 and E_2 are disjoint, then

$$\int_E f(x)\,dx \;=\; \int_{E_1} f(x)\,dx + \int_{E_2} f(x)\,dx$$

by using the Lebesgue integral as a limit of a sum.

Choosing subdivision points y_k such that

$$\alpha = y_0 < y_1 < \cdots < y_n = \beta$$

the set E is divided into subsets E_k while E_1 and E_2 are divided into disjoint subsets $E_k^{(1)}, E_k^{(2)}$ such that $E_k = E_k^{(1)} \cup E_k^{(2)}$ and

$$m(E_k) \;=\; m(E_k^{(1)}) + m(E_k^{(2)})$$

Then

$$\int_E f(x)\,dx \;=\; \lim \sum_{k=1}^{n} y_k\, m(E_k) \;=\; \lim \sum_{k=1}^{n} y_k\, m(E_k^{(1)}) + \lim \sum_{k=1}^{n} y_k\, m(E_k^{(2)})$$

$$=\; \int_{E_1} f(x)\,dx + \int_{E_2} f(x)\,dx$$

For another method see Problem 4.14.

4.14. Prove Theorem 4-6, page 55 without using the Lebesgue integral as the limit of a sum [see Problem 4.13].

Suppose that for $x \in E$, $\alpha < f(x) < \beta$ while for E_1 and E_2, $\alpha_1 < f(x) < \beta_1$ and $\alpha_2 < f(x) < \beta_2$ respectively. We can take α as the smaller of α_1, α_2 and β as the larger of β_1, β_2.

Let us choose as mode of subdivision of (α, β) the points $\alpha = y_0 < y_1 < \cdots < y_n = \beta$. Two cases can arise.

(1) Some of these are points of subdivision for (α_1, β_1) and the remaining ones are points of subdivision for (α_2, β_2).

(2) Two of the subdivision points do not happen to be the same as the larger of α_1, α_2 or the smaller of β_1, β_2.

If the second case arises we revise the points of subdivision so as to include the two extra points of subdivision. The refinement does not increase upper sums nor decrease lower sums [Problem 4.4].

Then if we denote upper sums corresponding to E, E_1, E_2 by S, S_1, S_2 respectively, we have

$$S \;\geqq\; S_1 + S_2$$

or using the fact that $f(x)$ is integrable on E, E_1 and E_2,

$$\int_E f(x)\,dx \;\geqq\; \int_{E_1} f(x)\,dx + \int_{E_2} f(x)\,dx \tag{1}$$

Similarly if we denote lower sums corresponding to E, E_1, E_2 by s, s_1, s_2, we find

$$s \;\leqq\; s_1 + s_2$$

or

$$\int_E f(x)\,dx \;\leqq\; \int_{E_1} f(x)\,dx + \int_{E_2} f(x)\,dx \tag{2}$$

From (1) and (2) we see that

$$\int_E f(x)\,dx \;=\; \int_{E_1} f(x)\,dx + \int_{E_2} f(x)\,dx$$

4.15. Prove Theorem 4-7, page 55: If $f(x)$ is bounded and measurable on $E = \bigcup_{k=1}^{\infty} E_k$ where the sets E_k are measurable and disjoint, then

$$\int_E f(x)\,dx \;=\; \int_{E_1} f(x)\,dx + \int_{E_2} f(x)\,dx + \cdots$$

Let $E = S_n \cup R_n$ where $S_n = \bigcup_{k=1}^{n} E_k$, $R_n = \bigcup_{k=n+1}^{\infty} E_k$. Then since $f(x)$ is integrable on

S_n and R_n, we have

$$\int_E f(x)\,dx \;=\; \int_{S_n} f(x)\,dx \;+\; \int_{R_n} f(x)\,dx$$

$$=\; \int_{E_1} f(x)\,dx \;+\; \cdots \;+\; \int_{E_n} f(x)\,dx \;+\; \int_{R_n} f(x)\,dx$$

Now by Theorem 4-5 [the mean-value theorem] since $|f(x)| \leqq M$,

$$\left| \int_{R_n} f(x)\,dx \right| \;\leqq\; M\,m(R_n) \tag{1}$$

But since $m(E) = m(S_n) + m(R_n) = m(E_1) + \cdots + m(E_n) + m(R_n)$ and since $m(E) = m(E_1) + m(E_2) + \cdots$ is finite, the infinite series converges so that $\lim\limits_{n \to \infty} m(R_n) = 0$. Thus taking the limit as $n \to \infty$ in (1), we find that

$$\lim_{n \to \infty} \int_{R_n} f(x)\,dx \;=\; 0 \tag{2}$$

Then

$$\int_E f(x)\,dx \;=\; \int_{E_1} f(x)\,dx \;+\; \int_{E_2} f(x)\,dx \;+\; \cdots$$

4.16. Prove that if $f(x)$ is bounded and measurable on E, and c is any constant, then

$$\int_E [f(x) + c]\,dx \;=\; \int_E f(x)\,dx \;+\; c\,m(E)$$

Suppose that the mode of subdivision of $\alpha < f(x) < \beta$ is made by the points $\alpha = y_0 < y_1 < \cdots < y_n = \beta$. Then the upper sum corresponding to $f(x)$ is

$$S \;=\; \sum_{k=1}^{n} y_k\,m(E_k) \tag{1}$$

where

$$E_k \;=\; \{x : x \in E,\; y_{k-1} \leqq f(x) < y_k\} \tag{2}$$

Similarly the upper sum corresponding to $f(x) + c$ is

$$S_1 \;=\; \sum_{k=1}^{n} (y_k + c)\,m(E_k) \;=\; \sum_{k=1}^{n} y_k\,m(E_k) \;+\; c\sum_{k=1}^{n} m(E_k)$$

i.e.

$$S_1 \;=\; S + c\,m(E)$$

Taking the limit as $n \to \infty$, we have

$$\int_E [f(x) + c]\,dx \;=\; \int_E f(x)\,dx \;+\; c\,m(E)$$

4.17. Prove Theorem 4-8, page 56: If $f(x)$ and $g(x)$ are bounded and measurable on E, then

$$\int_E [f(x) + g(x)]\,dx \;=\; \int_E f(x)\,dx \;+\; \int_E g(x)\,dx$$

Let $E = \bigcup\limits_{k=1}^{n} E_k$ where the E_k are given by (2) of Problem 4.16. Then

$$\int_E [f(x) + g(x)]\,dx \;=\; \sum_{k=1}^{n} \int_{E_k} [f(x) + g(x)]\,dx$$

$$\geqq\; \sum_{k=1}^{n} \int_{E_k} [y_{k-1} + g(x)]\,dx$$

$$=\; \sum_{k=1}^{n} y_{k-1}\,m(E_k) \;+\; \sum_{k=1}^{n} \int_{E_k} g(x)\,dx$$

$$=\; s \;+\; \int_E g(x)\,dx$$

Thus we have $$\int_E [f(x) + g(x)]\, dx \;\geqq\; \int_E f(x)\, dx \;+\; \int_E g(x)\, dx \tag{1}$$

Similarly
$$\int_E [f(x) + g(x)]\, dx \;=\; \sum_{k=1}^{n} \int_{E_k} [f(x) + g(x)]\, dx$$

$$\leqq\; \sum_{k=1}^{n} \int_{E_k} [y_k + g(x)]\, dx$$

$$=\; \sum_{k=1}^{n} y_k\, m(E_k) \;+\; \sum_{k=1}^{n} \int_{E_k} g(x)$$

$$=\; S \;+\; \int_E g(x)\, dx$$

Thus we have $$\int_E [f(x) + g(x)]\, dx \;\leqq\; \int_E f(x)\, dx \;+\; \int_E g(x)\, dx \tag{2}$$

From (1) and (2) we must have

$$\int_E [f(x) + g(x)]\, dx \;=\; \int_E f(x)\, dx \;+\; \int_E g(x)\, dx$$

The result is easily extended to any finite number of functions [see Problem 4.48].

4.18. Prove Theorem 4-10, page 56: If $f(x) \leqq g(x)$ on E, then

$$\int_E f(x)\, dx \;\leqq\; \int_E g(x)\, dx$$

Since $g(x) - f(x)$ is non-negative on E, i.e. $g(x) - f(x) \geqq 0$, we have

$$\int_E [g(x) - f(x)]\, dx \;\geqq\; 0$$

i.e. $$\int_E f(x)\, dx \;\leqq\; \int_E g(x)\, dx$$

4.19. Prove Theorem 4-11, page 56: If $f(x)$ is bounded and measurable on E, then $|f(x)|$ is Lebesgue integrable on E. Conversely if $|f(x)|$ is bounded and measurable on E, then $f(x)$ is Lebesgue integrable on E.

If $f(x)$ is bounded and measurable on E, then so also is $|f(x)|$ [see Theorem 3-13, page 44]. Thus $|f(x)|$ is Lebesgue integrable on E.

Similarly if $|f(x)|$ is bounded and measurable on E, then so also is $f(x)$. Thus $f(x)$ is Lebesgue integrable on E.

4.20. Prove Theorem 4-12, page 56: $\left| \int_E f(x)\, dx \right| \leqq \int_E |f(x)|\, dx$ under the conditions of Theorem 4-11.

Method 1.

We have $$-|f(x)| \;\leqq\; f(x) \;\leqq\; |f(x)|$$

Then by Problem 4.18 we have on integrating over the set E,

$$-\int_E |f(x)|\, dx \;\leqq\; \int_E f(x)\, dx \;\leqq\; \int_E |f(x)|\, dx$$

i.e. $$\left| \int_E f(x)\, dx \right| \;\leqq\; \int_E |f(x)|\, dx$$

Method 2.

If $E^+ = E[f(x) \geqq 0]$ and $E^- = E[f(x) < 0]$, then

$$\int_E f(x)\,dx \;=\; \int_{E^+} f(x)\,dx \;+\; \int_{E^-} f(x)\,dx \;=\; \int_{E^+} |f(x)|\,dx \;-\; \int_{E^-} |f(x)|\,dx$$

and

$$\int_E |f(x)|\,dx \;=\; \int_{E^+} |f(x)|\,dx \;+\; \int_{E^-} |f(x)|\,dx$$

Then

$$\left| \int_E f(x)\,dx \right| \;=\; \left| \int_{E^+} |f(x)|\,dx \;-\; \int_{E^-} |f(x)|\,dx \right|$$

$$\leqq \; \int_{E^+} |f(x)|\,dx \;+\; \int_{E^-} |f(x)|\,dx \;=\; \int_E |f(x)|\,dx$$

4.21. Prove Theorem 4-14, page 56: If $f(x) \geqq 0$ is bounded and measurable on E and $\int_E f(x)\,dx = 0$, then $f(x) = 0$ almost everywhere on E.

Method 1.

Since $f(x)$ is bounded, there is a constant M such that $0 \leqq f(x) \leqq M$. Consider the sets

$$E_1 = \{x : f(x) = 0\}, \quad E_2 = \left\{ x : \frac{M}{2} < f(x) \leqq M \right\}, \quad E_3 = \left\{ x : \frac{M}{3} < f(x) \leqq \frac{M}{2} \right\}$$

and in general

$$E_k = \left\{ x : \frac{M}{k} < f(x) \leqq \frac{M}{k-1} \right\}, \qquad k = 2, 3, \ldots$$

Since the sets are measurable and disjoint,

$$m(E) \;=\; m(E_1) + m(E_2) + \cdots$$

where $E = \bigcup_{k=1}^{\infty} E_k$. Now by the mean-value theorem,

$$\frac{M}{k}\, m(E_k) \;\leqq\; \int_{E_k} f(x)\,dx \;\leqq\; \frac{M}{k-1}\, m(E_k), \qquad k = 2, 3, \ldots$$

From the inequality on the left and the fact that the integral over E_k cannot exceed the integral over E, we have

$$m(E_k) \;\leqq\; \frac{k}{M} \int_{E_k} f(x)\,dx \;\leqq\; \frac{k}{M} \int_E f(x)\,dx \;=\; 0$$

Then $m(E_k) = 0$ for $k = 2, 3, \ldots$. It follows that

$$m(E_2 \cup E_3 \cup \cdots) \;=\; m(E_2) + m(E_3) + \cdots \;=\; 0$$

i.e. the measure of the set for which $f(x) \geqq 0$ is zero, so that $f(x) = 0$ almost everywhere.

Method 2.

Suppose that contained in E there is a set A of positive measure for which $f(x) > 0$. Then we can express A as the union of mutually disjoint measurable sets

$$A_1 = \{x : f(x) > 1\}, \quad A_2 = \{x : f(x) > \tfrac{1}{2}\}, \quad A_3 = \{x : f(x) > \tfrac{1}{3}\} \cdots$$

Since $m(A) = m(A_1) + m(A_2) + \cdots$ and $m(A) > 0$, it follows that there is some set A_k such that $m(A_k)$ is a positive number, say p.

Then by the mean-value theorem and the fact that the integral over A_k cannot exceed the integral over E, we have

$$\int_E f(x)\,dx \;\geqq\; \int_{A_k} f(x)\,dx \;\geqq\; \frac{m(A_k)}{k} \;=\; \frac{p}{k} \;>\; 0$$

which contradicts the hypothesis that the integral over E is zero. Thus the set A cannot be of positive measure and it follows that $f(x) = 0$ almost everywhere.

LEBESGUE'S THEOREM ON BOUNDED CONVERGENCE

4.22. Prove Lebesgue's theorem on bounded convergence [Theorem 4-15, page 56]: Let $\langle f_n(x) \rangle$ be a sequence of functions measurable on E such that $\lim_{n \to \infty} f_n(x) = f(x)$. Then if $|f_n(x)| \leq M$,

$$\lim_{n \to \infty} \int_E f_n(x) \, dx = \int_E f(x) \, dx$$

From the fact that $|f_n(x)| \leq M$ and $\lim_{n \to \infty} f_n(x) = f(x)$, it follows that $|f(x)| \leq M$. Then $f(x)$ is bounded and measurable and thus integrable. The result to be proved is equivalent to proving that

$$\lim_{n \to \infty} \int_E [f(x) - f_n(x)] \, dx = 0 \tag{1}$$

or since

$$\left| \int_E [f(x) - f_n(x)] \, dx \right| \leq \int_E |f(x) - f_n(x)| \, dx$$

it will follow if we can prove that

$$\lim_{n \to \infty} \int_E |f(x) - f_n(x)| \, dx = 0 \tag{2}$$

Let E be represented as the union of the disjoint measurable sets

$$E_1 = \{x : |f - f_1| < \epsilon, \ |f - f_2| < \epsilon, \ldots\}$$

$$E_2 = \{x : |f - f_1| \geq \epsilon, \ |f - f_2| < \epsilon, \ldots\}$$

$$E_3 = \{x : |f - f_2| \geq \epsilon, \ |f - f_3| < \epsilon, \ldots\}$$

$$\cdots\cdots\cdots\cdots\cdots\cdots\cdots\cdots\cdots\cdots\cdots\cdots\cdots\cdots\cdots$$

$$E_n = \{x : |f - f_{n-1}| \geq \epsilon, \ |f - f_n| < \epsilon, \ldots\}$$

etc., i.e.

$$E = \bigcup_{k=1}^{\infty} E_k$$

Then if $S_n = \bigcup_{k=1}^{n} E_k$ and $R_n = \bigcup_{k=n+1}^{\infty} E_k$, we have

$$\int_E |f(x) - f_n(x)| \, dx = \int_{S_n} |f(x) - f_n(x)| \, dx + \int_{R_n} |f(x) - f_n(x)| \, dx \tag{3}$$

Now on S_n we have $|f(x) - f_n(x)| < \epsilon$. Also since $|f_n(x)| \leq M$ and $|f(x)| \leq M$, we have on R_n, $|f(x) - f_n(x)| \leq |f(x)| + |f_n(x)| \leq 2M$. Thus (3) becomes on using the mean-value theorem,

$$\int_E |f(x) - f_n(x)| \, dx \leq \epsilon \, m(S_n) + 2M \, m(R_n) \tag{4}$$

Since $\lim_{n \to \infty} m(S_n) = m(E)$ and $\lim_{n \to \infty} m(R_n) = 0$, we have from (4)

$$\overline{\lim_{n \to \infty}} \int_E |f(x) - f_n(x)| \, dx \leq \epsilon \, m(E)$$

or letting $\epsilon \to 0$,

$$\overline{\lim_{n \to \infty}} \int_E |f(x) - f_n(x)| \, dx = 0$$

i.e.

$$\lim_{n \to \infty} \int_E |f(x) - f_n(x)| \, dx = 0$$

which proves the required result.

4.23. Prove that the conditions in Problem 4.22 can be relaxed so that they hold almost everywhere.

This follows at once since sets of measure zero do not affect the values of the integrals of $f(x)$ and $f_n(x)$ on E.

4.24. Give an example of a sequence of functions $\langle f_n(x) \rangle$ such that $\lim\limits_{n \to \infty} f_n(x) = f(x)$ and having the property that

$$\lim_{n \to \infty} \int_a^b f_n(x)\,dx \;=\; \int_a^b \lim_{n \to \infty} f_n(x)\,dx$$

$$\lim_{n \to \infty} (\mathcal{R}) \int_a^b f_n(x)\,dx \;\neq\; (\mathcal{R}) \int_a^b \lim_{n \to \infty} f_n(x)\,dx$$

Let $\langle r_k \rangle$ denote a sequence representing all the rational numbers in $[0,1]$. [This is possible since the set of rational numbers in $[0,1]$ is denumerable]. Define a sequence of functions $\langle f_n(x) \rangle$ in $[0,1]$ such that

$$f_n(x) \;=\; \begin{cases} 1 & x = r_1, r_2, \ldots, r_n \\ 0 & x \neq r_1, r_2, \ldots, r_n \end{cases}$$

Then

$$\lim_{n \to \infty} f_n(x) \;=\; f(x) \;=\; \begin{cases} 1 & x \text{ is a rational number in } [0,1] \\ 0 & x \text{ is an irrational number in } [0,1] \end{cases}$$

Now from the definition of $f_n(x)$ we have

$$\int_0^1 f_n(x)\,dx \;=\; 0 \quad \text{and so} \quad \lim_{n \to \infty} \int_0^1 f_n(x)\,dx \;=\; 0$$

Also using Problem 4.9,

$$\int_0^1 \lim_{n \to \infty} f_n(x) \;=\; \int_0^1 f(x)\,dx \;=\; 0$$

Thus

$$\lim_{n \to \infty} \int_0^1 f_n(x)\,dx \;=\; \int_0^1 \lim_{n \to \infty} f_n(x)\,dx$$

We would expect this of course in view of Problem 4.22.

However, if we use the Riemann integral we have

$$(\mathcal{R}) \int_0^1 f_n(x)\,dx \;=\; 0 \quad \text{so that} \quad \lim_{n \to \infty} (\mathcal{R}) \int_0^1 f_n(x)\,dx \;=\; 0$$

but from Problem A.2, page 159,

$$(\mathcal{R}) \int_0^1 \lim_{n \to \infty} f_n(x)\,dx \;=\; (\mathcal{R}) \int_0^1 f(x)\,dx$$

does not exist. Thus

$$\lim_{n \to \infty} (\mathcal{R}) \int_0^1 f_n(x)\,dx \;\neq\; (\mathcal{R}) \int_0^1 \lim_{n \to \infty} f_n(x)\,dx$$

RELATIONSHIP OF RIEMANN AND LEBESGUE INTEGRALS

4.25. Prove that if $f(x)$ is Riemann integrable in $[a, b]$, then $f(x)$ must be measurable.

Suppose that the interval $[a, b]$ is subdivided into n parts by the subdivision points

$$a = x_0 < x_1 < x_2 < \cdots < x_n = b \tag{1}$$

Consider the functions

$$\psi_n(x) = m_k^{(n)}, \qquad \Psi_n(x) = M_k^{(n)}, \qquad x_k < x \leq x_{k+1} \tag{2}$$

where $m_k^{(n)}$ and $M_k^{(n)}$ are the g.l.b. and l.u.b. of $f(x)$ in the interval $x_k < x \leq x_{k+1}$. We have used the superscript n to emphasize the fact that there are n subdivisions.

Integrating the functions in (2) from a to b, we have

$$\int_a^b \psi_n(x)\,dx \;=\; \sum_{k=1}^n m_k^{(n)}(x_k - x_{k-1}) \tag{3}$$

$$\int_a^b \Psi_n(x)\,dx \;=\; \sum_{k=1}^n M_k^{(n)}(x_k - x_{k-1}) \tag{4}$$

These integrals are respectively the lower and upper Riemann sums of $f(x)$. It follows since $f(x)$ is Riemann integrable, that

$$\lim_{n \to \infty} \int_a^b \psi_n(x)\, dx \;=\; \lim_{n \to \infty} \int_a^b \Psi_n(x)\, dx \;=\; (\mathcal{R}) \int_a^b f(x)\, dx \qquad (5)$$

Now as $n \to \infty$ the functions $\psi_n(x)$ and $\Psi_n(x)$ converge to the functions $\psi(x)$ and $\Psi(x)$ respectively where $\psi(x)$ represents the greatest lower bound of $f(x)$ in a δ neighborhood of x as $\delta \to 0$ while $\Psi(x)$ represents the least upper bound of $f(x)$ in a δ neighborhood of x as $\delta \to 0$.

Since $\psi_n(x)$ and $\Psi_n(x)$ are measurable functions, the limits of these are also measurable [Theorem 3-19, page 44] so that $\psi(x)$ and $\Psi(x)$ are measurable. By Lebesgue's convergence theorem and (5) we have

$$\lim_{n \to \infty} \int_a^b \psi_n(x)\, dx \;=\; \int_a^b \psi(x)\, dx \;=\; (\mathcal{R}) \int_a^b f(x)\, dx \qquad (6)$$

$$\lim_{n \to \infty} \int_a^b \Psi_n(x)\, dx \;=\; \int_a^b \Psi(x)\, dx \;=\; (\mathcal{R}) \int_a^b f(x)\, dx \qquad (7)$$

so that on subtraction,
$$\int_a^b [\Psi(x) - \psi(x)]\, dx \;=\; 0 \qquad (8)$$

But since $\Psi(x) \geqq \psi(x)$, it follows from Problem 4.21 that $\Psi(x) = \psi(x)$ almost everywhere in $[a, b]$. Now it is clear that $\psi(x) \leqq f(x) \leqq \Psi(x)$, from which we see that $f(x) = \psi(x)$ almost everywhere and that $f(x)$ is measurable since $\psi(x)$ is.

4.26. Prove Theorem 4-17, page 57: If $f(x)$ is Riemann integrable in $[a, b]$, then (a) it is Lebesgue integrable in $[a, b]$ and (b) the Riemann and Lebesgue integrals of $f(x)$ in $[a, b]$ are equal.

Using the notation of Problem 4.25, we have
$$\psi(x) \;\leqq\; f(x) \;\leqq\; \Psi(x)$$
so that by (6) and (7) of that problem,

$$(\mathcal{R}) \int_a^b f(x)\, dx \;=\; \int_a^b \psi(x)\, dx \;\leqq\; \int_a^b f(x)\, dx \;\leqq\; \int_a^b \Psi(x)\, dx \;=\; (\mathcal{R}) \int_a^b f(x)\, dx$$

Thus
$$\int_a^b f(x)\, dx \;=\; (\mathcal{R}) \int_a^b f(x)\, dx$$

and the required results are proved.

4.27. Prove Theorem 4-18, page 57: A function $f(x)$ is Riemann integrable in $[a, b]$ if and only if the set of discontinuities of $f(x)$ has measure zero, i.e. $f(x)$ is continuous almost everywhere.

If $f(x)$ is continuous at a point x, then using the notation of Problem 4.25 we must have $\psi(x) = \Psi(x)$. Conversely if $\psi(x) = \Psi(x)$ at x, then $f(x)$ is continuous at x. In other words $f(x)$ is continuous at x if and only if $\psi(x) = \Psi(x)$ at x.

Now from the results of Problem 4.26 we see that $f(x)$ is Riemann integrable in $[a, b]$ if and only if [see equations (5), (6) and (7) of Problem 4.25].

$$\int_a^b \psi(x)\, dx \;=\; \lim_{n \to \infty} \int_a^b \psi_n(x)\, dx \;=\; \lim_{n \to \infty} \int_a^b \Psi_n(x)\, dx \;=\; \int_a^b \Psi(x)\, dx$$

or $\psi(x) = \Psi(x)$ almost everywhere, i.e. $f(x)$ is continuous almost everywhere.

4.28. Prove Theorem 4-19, page 57: If $f(x)$ is continuous almost everywhere in $[a, b]$, then it is Lebesgue integrable in $[a, b]$.

This follows at once from Problems 4.26 and 4.27.

Supplementary Problems

DEFINITION OF THE LEBESGUE INTEGRAL FOR BOUNDED MEASURABLE FUNCTIONS

4.29. Use the definition of the Lebesgue integral to show that if c is any constant

$$\int_a^b c\,dx = c(b-a)$$

4.30. Evaluate $\int_0^5 f(x)\,dx$ if $f(x) = \begin{cases} -1 & 0 < x < 3 \\ 4 & 3 < x < 5 \end{cases}$ by using the definition.

4.31. Work Problem 4.30 if $f(x) = \begin{cases} 2 & 0 \leqq x < 1 \\ 3 & 2 < x \leqq 4 \\ -1 & 4 < x \leqq 5 \end{cases}$.

4.32. Prove that in any refinement of a partition or mode of subdivision the lower sums cannot decrease [see Problem 4.4].

4.33. Give a proof of the theorem in Problem 4.6 without referring to a diagram.

4.34. Show how to work Problem 4.9 by not referring to a diagram.

4.35. Let $f(x) = \begin{cases} 1 & \text{if } x \text{ is an irrational number in } (-4, 4) \\ -2 & \text{if } x \text{ is a rational number in } (-4, 4) \end{cases}$. Evaluate $\int_{-4}^4 f(x)\,dx$.

4.36. If Q is the set of all rational numbers and $f(x) = 3$, find $\int_Q f(x)\,dx$.

4.37. If K is the Cantor set [page 5] and $f(x) = 2$ on this set, evaluate $\int_K f(x)\,dx$.

4.38. Prove that the definition of the Lebesgue integral on page 53 is independent of the choice of α and β so long as $\alpha < f(x) < \beta$.

4.39. Prove the converse of the result in Problem 4.11.

THEOREMS INVOLVING THE LEBESGUE INTEGRAL

4.40. Prove Theorem 4-2, page 55.

4.41. Prove Theorem 4-3, page 55.

4.42. Prove that if E_1 and E_2 are measurable disjoint sets and $E = E_1 \cup E_2$, then if $f(x) = \begin{cases} c_1, & x \in E_1 \\ c_2, & x \in E_2 \end{cases}$ where c_1 and c_2 are given constants,

$$\int_E f(x)\,dx = c_1\,m(E_1) + c_2\,m(E_2)$$

4.43. Generalize the result of Problem 4.42.

4.44. Use Problems 4.42 or 4.43 to work (a) Problem 4.9, (b) Problem 4.30, (c) Problem 4.31, (d) Problem 4.35.

4.45. Prove Theorem 4-4, page 55.

4.46. Prove or disprove: If $f(x)$ is continuous on a measurable set E, then there exists a number ξ such that

$$\int_E f(x)\,dx = f(\xi)\,m(E)$$

4.47. Extend the proof of Theorem 4-6, page 55, [see Problem 4.13] to any finite union of sets $E_1, E_2, \ldots, E_n$.

4.48. Extend the proof of Theorem 4-8, page 56, [see Problem 4.17] to any finite number of functions.

4.49. Prove that the functions (a) x, (b) x^2, (c) $3x^2 + 4x - 2$ are Lebesgue integrable in $[0, 1]$. Can you find the value of the integrals? Explain.

4.50. Verify that the sets E_k of Problem 4.16 are measurable and disjoint and that if $E = \cup E_k$, then $m(E) = \sum m(E_k)$.

4.51. Prove the result of Problem 4.16 without using the Lebesgue integrals as a limit of a sum.

4.52. If $f(x) \geqq 0$ is bounded and measurable on E, prove that $\displaystyle\int_E f(x)\, dx \geqq 0$. Can the equality sign be removed if $f(x) > 0$? Explain.

4.53. Prove that if $f(x)$ and $g(x)$ are bounded and measurable on E, then
$$\int_E [f(x) - g(x)]\, dx = \int_E f(x)\, dx - \int_E g(x)\, dx$$

4.54. Prove that
$$\int_{E_1 \cup E_2} f(x)\, dx = \int_{E_1} f(x)\, dx + \int_{E_2} f(x)\, dx - \int_{E_1 \cap E_2} f(x)\, dx$$

4.55. Generalize the result of Problem 4.54 to three or more sets $E_1, E_2, E_3, \ldots$.

4.56. If $\langle f_k(x) \rangle$, $k = 1, 2, \ldots, n$ are bounded and measurable on E and c_k, $k = 1, 2, \ldots, n$, are any constants, prove that
$$\int_E \left[\sum_{k=1}^n c_k f_k(x) \right] dx = \sum_{k=1}^n c_k \int_E f_k(x)\, dx$$

4.57. Prove Theorem 4-9, page 56.

4.58. Prove that
$$\left| \int_E f(x)\, g(x)\, dx \right| \leqq \frac{1}{2} \left[\int_E |f(x)|^2\, dx + \int_E |g(x)|^2\, dx \right]$$
if $f(x)$ and $g(x)$ are bounded and measurable on E.

4.59. Prove Theorem 4-13, page 56.

4.60. If $f_1(x)$ and $f_2(x)$ are bounded and measurable in (a, b) and
$$\int_a^b [f_1(x) - f_2(x)]^2\, dx = 0$$
prove that $f_1(x) = f_2(x)$ almost everywhere in (a, b).

4.61. State and prove a theorem analogous to that of Problem 4.28 if the interval $[a, b]$ is replaced by a set E.

4.62. Evaluate $\displaystyle\int_0^\pi \sin x\, dx$.   *Ans.* 2

4.63.   Evaluate $\displaystyle\int_{-1}^{1} \frac{dx}{x^2 + 1}$.   *Ans.* $\pi/2$

4.64.   Let $f(x)$ be bounded and integrable on E. Prove that given $\epsilon > 0$, there exists $\delta > 0$ such that if A is any measurable set contained in E then

$$\int_A |f(x)|\, dx \;<\; \epsilon \quad \text{whenever} \quad m(A) < \delta$$

LEBESGUE'S THEOREM ON BOUNDED CONVERGENCE

4.65.   Evaluate $\displaystyle\lim_{n \to \infty} \int_0^1 e^{-nx^2}\, dx$.   *Ans.* 0

4.66.   Evaluate   (a) $\displaystyle\lim_{n \to \infty} \int_0^1 \frac{dx}{(1 + x^2)^n}$ (b) $\displaystyle\lim_{n \to \infty} \int_0^1 \frac{dx}{(1 + x/n)^n}$.

4.67.   Prove Theorem 4-16, page 56.

4.68.   Illustrate Theorem 4-16 by means of an example.

4.69.   Prove that if $f_k(x) \geqq 0$ and $\displaystyle\sum_{k=1}^{\infty} \int_a^b f_k(x)\, dx$ converges, then $\displaystyle\sum_{k=1}^{\infty} f_k(x)$ converges almost everywhere.

4.70.   Prove Lebesgue's theorem on bounded convergence by using Egorov's theorem, page 45.

Chapter 5

The Lebesgue Integral for Unbounded Functions

THE LEBESGUE INTEGRAL FOR NON-NEGATIVE UNBOUNDED FUNCTIONS

In Chapter 4 we defined the Lebesgue integral and obtained its properties for bounded measurable functions. In this chapter we extend these results to unbounded measurable functions.

Consider first the case where $f(x) \geqq 0$ is unbounded and measurable. If p is a natural number we shall use the notation

$$[f(x)]_p \quad = \quad \begin{cases} f(x) & \text{for all } x \in E \text{ such that } f(x) \leqq p \\ p & \text{for all } x \in E \text{ such that } f(x) > p \end{cases} \tag{1}$$

Then for each p, $[f(x)]_p$ is bounded and measurable and thus Lebesgue integrable. We define the Lebesgue integral of $f(x)$ on E as

$$\int_E f(x)\,dx \quad = \quad \lim_{p \to \infty} \int_E [f(x)]_p\,dx \tag{2}$$

The limit is either a non-negative number or is infinite. If the limit is a non-negative number, then the Lebesgue integral of $f(x)$ exists and is equal to this number and we say that $f(x)$ is *Lebesgue integrable* or briefly *integrable* on E. If the limit is infinite we say that $f(x)$ is not Lebesgue integrable on E or that the integral does not exist.

THE LEBESGUE INTEGRAL FOR ARBITRARY UNBOUNDED FUNCTIONS

If $f(x) \leqq 0$ we can define the Lebesgue integral of $f(x)$ as

$$\int_E f(x)\,dx \quad = \quad -\int_E |f(x)|\,dx \tag{3}$$

where the integral on the right is obtained as above since $|f(x)| \geqq 0$.

In the general case where $f(x)$ may have arbitrary sign, we let

$$f^+(x) \quad = \quad \begin{cases} f(x) & \text{for all } x \in E \text{ such that } f(x) \geqq 0 \\ 0 & \text{for all } x \in E \text{ such that } f(x) < 0 \end{cases} \tag{4}$$

$$f^-(x) \quad = \quad \begin{cases} 0 & \text{for all } x \in E \text{ such that } f(x) \geqq 0 \\ -f(x) & \text{for all } x \in E \text{ such that } f(x) < 0 \end{cases} \tag{5}$$

where it is to be noted that $f^+(x)$ and $f^-(x)$ are both non-negative. Then it follows that

$$f(x) \quad = \quad f^+(x) - f^-(x) \tag{6}$$

This leads us to define the integral of $f(x)$ over E as the difference of the integrals of non-negative functions, i.e.

$$\int_E f(x)\,dx \quad = \quad \int_E f^+(x)\,dx - \int_E f^-(x)\,dx \tag{7}$$

The integral on the left of (7) will certainly exist when the two integrals on the right exist and in such case we will say that $f(x)$ is *Lebesgue integrable* or briefly *integrable* on E.

It should be mentioned that in case $f(x)$ is bounded the definition given by (7) reduces to that given in Chapter 4.   [See Problem 5.43.]

THEOREMS INVOLVING LEBESGUE INTEGRALS OF UNBOUNDED FUNCTIONS

Most of the theorems of Chapter 4 involving Lebesgue integrals of bounded measurable functions can be extended to the case of unbounded measurable functions.   In general, proofs of such extended theorems are achieved by first using a limiting procedure as in (2), page 72, for the case where the functions are non-negative and then using the definition (7), page 72, to obtain the case where the functions are arbitrary in sign.

In the following we list some important theorems many of which are related to those in Chapter 4.   Unless otherwise stated we assume that all sets and functions are measurable and that the functions may have arbitrary sign and may be unbounded.

Theorem 5-1. If $f(x) \geqq 0$, then $\displaystyle\int_E f(x)\,dx$ exists if and only if $\displaystyle\int_E [f(x)]_p\,dx$ is uniformly bounded.

Theorem 5-2. If $|f(x)| \leqq g(x)$ where $g(x)$ is integrable on E, then $f(x)$ is also integrable on E and
$$\int_E |f(x)|\,dx \;\leqq\; \int_E g(x)\,dx$$
The result is also true if $|f(x)| \leqq g(x)$ almost everywhere on E.

Theorem 5-3. A function $f(x)$ is integrable on E if and only if $|f(x)|$ is integrable on E and in such case
$$\left| \int_E f(x)\,dx \right| \;\leqq\; \int_E |f(x)|\,dx$$
Because of this we say that $f(x)$ is integrable on E if and only if it is *absolutely integrable* on E.

Theorem 5-4. If $\displaystyle\int_E f(x)\,dx$ exists, then $f(x)$ is finite almost everywhere in E.

Theorem 5-5. If E has measure zero, then
$$\int_E f(x)\,dx \;=\; 0$$

Theorem 5-6. If $\displaystyle\int_E f(x)\,dx$ exists and if A is a measurable subset of E, then $\displaystyle\int_A f(x)\,dx$ also exists.   In such case we have
$$\int_A |f(x)|\,dx \;\leqq\; \int_E |f(x)|\,dx$$

Theorem 5-7. If $E = E_1 \cup E_2$ where E_1 and E_2 are disjoint, then
$$\int_E f(x)\,dx \;=\; \int_{E_1} f(x)\,dx \;+\; \int_{E_2} f(x)\,dx$$
The result is easily generalized to finite unions of sets.

Theorem 5-8. If $E = E_1 \cup E_2 \cup \cdots$ where $E_1, E_2, \ldots$ are mutually disjoint, then if $\int_E f(x)\, dx$ exists

$$\int_E f(x)\, dx = \int_{E_1} f(x)\, dx + \int_{E_2} f(x)\, dx + \cdots$$

The result is not necessarily true if $f(x)$ is integrable on $E_1, E_2, \ldots$ without specific mention of its integrability on E [see Problem 5.16].

Theorem 5-9. $$\int_E [f(x) + g(x)]\, dx = \int_E f(x)\, dx + \int_E g(x)\, dx$$

The result is easily generalized to any finite number of functions.

Theorem 5-10. $\int_E c\, f(x)\, dx = c \int_E f(x)\, dx$ for any constant c.

The result is easily generalized to any finite number of functions.

Theorem 5-11. If $f(x)$ is integrable on E and $g(x)$ is bounded, then $f(x)\, g(x)$ is integrable on E.

Theorem 5-12. If $f(x) = g(x)$ almost everywhere on E, then

$$\int_E f(x)\, dx = \int_E g(x)\, dx$$

and the existence of one integral implies the existence of the other.

Theorem 5-13. If $f(x) \geqq 0$ and $\int_E f(x)\, dx = 0$, then $f(x) = 0$ almost everywhere on E.

The result is also true if $f(x) \geqq 0$ almost everywhere on E.

Theorem 5-14. If $f(x)$ is integrable on E, then given $\epsilon > 0$ there exist a $\delta > 0$ and a set $A \subset E$ such that if $m(A) < \delta$

$$\left| \int_A f(x)\, dx \right| < \delta$$

Theorem 5-15. Let $f(x)$ be integrable in E. If $\langle E_k \rangle$ is a sequence of sets contained in E such that $\lim_{k \to \infty} m(E_k) = 0$, then

$$\lim_{k \to \infty} \int_{E_k} f(x)\, dx = 0$$

LEBESGUE'S DOMINATED CONVERGENCE THEOREM

The following theorem is a generalization of the Lebesgue bounded convergence theorem [Theorem 4-15] on page 56.

Theorem 5-16 (Dominated convergence). Let $\langle f_n(x) \rangle$ be a sequence of functions measurable on E such that $\lim_{n \to \infty} f_n(x) = f(x)$. Then if there exists a function $M(x)$ integrable on E such that $|f_n(x)| \leqq M(x)$ for all n, we have

$$\lim_{n \to \infty} \int_E f_n(x)\, dx = \int_E \lim_{n \to \infty} f_n(x)\, dx = \int_E f(x)\, dx \tag{7}$$

Because sets of measure zero can be omitted in the integrals, the conditions of the theorem can be relaxed so that they hold almost everywhere.

THE DOMINATED CONVERGENCE THEOREM FOR INFINITE SERIES

The Lebesgue theorem on dominated convergence stated above can be restated in terms of series as follows.

Theorem 5-17.   If $u_k(x), k = 1, 2, \ldots$, are measurable on E and if there exists an integrable function $M(x)$ on E such that $|s_n(x)| \leqq M(x)$ where $s_n(x) = \sum\limits_{k=1}^{n} u_k(x)$ and if $\lim\limits_{n \to \infty} s_n(x) = s(x)$, then

$$\int_E \left\{ \sum_{k=1}^{\infty} u_k(x) \right\} dx \;=\; \sum_{k=1}^{\infty} \int_E u_k(x)\, dx \qquad (8)$$

We also have the following theorem which is often useful.

Theorem 5-18.   If $\sum\limits_{k=1}^{\infty} u_k(x)$ is such that $|s_n(x)| = \left| \sum\limits_{k=1}^{n} u_k(x) \right| \leqq M$ for some constant M and if $v(x)$ is bounded and measurable on E, then

$$\int_E \left\{ v(x) \sum_{k=1}^{\infty} u_k(x) \right\} dx \;=\; \sum_{k=1}^{\infty} \int_E v(x)\, u_k(x)\, dx \qquad (9)$$

FATOU'S THEOREM

Theorem 5-19   [Fatou].   Let $\langle f_n(x) \rangle$ be a sequence of non-negative measurable functions defined on E and suppose that $\lim\limits_{n \to \infty} f_n(x) = f(x)$.   Then

$$\varliminf_{n \to \infty} \int_E f_n(x)\, dx \;\geqq\; \int_E f(x)\, dx \qquad (10)$$

Note that the theorem is also true if $\lim\limits_{n \to \infty} f_n(x) = f(x)$ almost everywhere.

If $f(x)$ is not integrable, the left hand side of (10) is infinite.

THE MONOTONE CONVERGENCE THEOREM

Theorem 5-20.   Let $\langle f_n(x) \rangle$ be a sequence of non-negative monotonic increasing functions on a set E and suppose that the sequence converges to $f(x)$.   Then

$$\lim_{n \to \infty} \int_E f_n(x)\, dx \;=\; \int_E f(x)\, dx \qquad (11)$$

provided that either side is finite.

The theorem can also be stated as a theorem on series as follows

Theorem 5-21.   Let $u_k(x) \geqq 0$.   Then

$$\int_E \left\{ \sum_{k=1}^{\infty} u_k(x) \right\} dx \;=\; \sum_{k=1}^{\infty} \int_E u_k(x)\, dx \qquad (12)$$

provided that either side converges.

APPROXIMATION OF INTEGRABLE FUNCTIONS BY CONTINUOUS FUNCTIONS

Although every continuous function in $[a, b]$ is integrable, an integrable function need not be continuous.   However, an integrable function can be approximated by continuous functions.   A fundamental theorem in this connection is

Theorem 5-22.   Let $f(x)$ be integrable in $[a, b]$.   Then given any $\epsilon > 0$, there exists a continuous function $g(x)$ in $[a, b]$ such that

$$\int_a^b |f(x) - g(x)|\, dx \;<\; \epsilon \qquad (13)$$

An important theorem which is related to Theorem 5-22 is the following.

Theorem 5-23. Let $f(x)$ be integrable in $[a, b]$. Then

$$\lim_{h \to 0} \int_a^b |f(x + h) - f(x)| \, dx \ = \ 0 \tag{14}$$

THE LEBESGUE INTEGRAL ON UNBOUNDED SETS OR INTERVALS

Up to now we have considered the Lebesgue integral on bounded sets. In order to extend the definition to unbounded sets such as the intervals $(a, \infty), (-\infty, b)$ or $(-\infty, \infty)$, we need only adopt a suitable limiting case of bounded sets.

Let us illustrate by defining the Lebesgue integrable on an unbounded set such as (a, ∞). Suppose first that $f(x)$ is non-negative, i.e. $f(x) \geqq 0$, and that $f(x)$ is Lebesgue integrable in (a, b) for all finite values of b. Then we define

$$\int_a^\infty f(x) \, dx \ = \ \lim_{b \to \infty} \int_a^b f(x) \, dx \tag{15}$$

and say that $f(x)$ is *Lebesgue integrable*, or briefly *integrable*, on (a, ∞) if the limit in (15) exists.

If $f(x)$ has arbitrary sign, then we define

$$\int_a^\infty f(x) \, dx \ = \ \int_a^\infty f^+(x) \, dx \ - \ \int_a^\infty f^-(x) \, dx \tag{16}$$

where $f^+(x), f^-(x)$ are non-negative and say that $f(x)$ is integrable on (a, ∞) if each of the integrals on the right of (16) exist in accordance with the definition (15).

Obvious extensions can be made for the intervals $(-\infty, b)$ and $(-\infty, \infty)$. In the last one, for example, we consider (a, b) and then let $a \to -\infty, b \to \infty$. If the set E is not an infinite interval but is nevertheless unbounded, we can define

$$\int_E f(x) \, dx \ = \ \lim_{\substack{a \to -\infty \\ b \to \infty}} \int_{E \cap (a, b)} f(x) \, dx \tag{17}$$

THEOREMS FOR LEBESGUE INTEGRALS ON UNBOUNDED SETS

Most of the theorems which have been proved involving Lebesgue integrals on bounded sets are also valid for unbounded sets. For example, we have [compare Theorem 5-3, page 73].

Theorem 5-24. A function $f(x)$ is integrable on E if and only if $|f(x)|$ is integrable on E regardless of whether E is bounded or unbounded and in such case we have

$$\left| \int_E f(x) \, dx \right| \ \leqq \ \int_E |f(x)| \, dx \tag{18}$$

Thus $f(x)$ is integrable on E if and only if $f(x)$ is *absolutely integrable* on E.

Similarly we can show that the other theorems proved for bounded sets, such as the Lebesgue dominated convergence theorem and Fatou's theorem, are also valid in case the sets are unbounded.

COMPARISON WITH THE RIEMANN INTEGRAL FOR UNBOUNDED SETS

Unlike the Lebesgue integral, the Riemann integral of $f(x)$ on an unbounded set E can exist even though the Riemann integral of $|f(x)|$ does not exist on E.   For example, although

$$\int_0^\infty \frac{\sin x}{x}\, dx = \lim_{b \to \infty} \int_0^b \frac{\sin x}{x}\, dx$$

exists as an (improper) Riemann integral, the integral

$$\int_0^\infty \left| \frac{\sin x}{x} \right| dx$$

does not exist [see Problems A.32, page 170, and A.33, page 171].   Thus the Riemann integral may exist when the Lebesgue integral does not exist.   However, we have the following important theorem.

Theorem 5-25.   If $|f(x)|$ is Riemann integrable on E, then $f(x)$ is both Riemann and Lebesgue integrable on E and the two integrals are equal.

Solved Problems

DEFINITION OF THE LEBESGUE INTEGRAL FOR NON-NEGATIVE UNBOUNDED FUNCTIONS

5.1.   Show that the function $[f(x)]_p$ defined on page 72 can be written as $\min\{f(x), p\}$ where min denotes minimum.

We have

$$[f(x)]_p \;=\; \begin{cases} f(x) & \text{for all } x \in E \text{ such that } f(x) \leq p \\ p & \text{for all } x \in E \text{ such that } f(x) > p \end{cases}$$

Then if $f(x)$ is less than or equal to p, $[f(x)]_p = f(x) = \min\{f(x), p\}$ while if $f(x)$ is greater than p, $[f(x)]_p = p = \min\{f(x), p\}$.

In either case $[f(x)]_p = \min\{f(x), p\}$.

5.2.   Show that $\displaystyle\lim_{p \to \infty} [f(x)]_p = f(x)$.

This follows at once from Problem 5.1 since $\min\{f(x), \infty\} = f(x)$.

5.3.   Show that $[f(x)]_p$ is bounded and measurable, and thus integrable for each p if $f(x)$ is measurable.

From Problem 5.1 we see that $[f(x)]_p \leq p$ and is thus bounded for each p.

Also since $[f(x)]_p = \min\{f(x), p\}$ where $f(x)$ and p are measurable, it follows from Problem 3.18, page 48, that $[f(x)]_p$ is measurable.

The last result can also be shown directly from the definition of a measurable function [see Problem 5.39].

5.4. Prove that $\int_0^8 \dfrac{dx}{\sqrt[3]{x}}$ exists and find its value.

Define the function

$$[f(x)]_p = \begin{cases} 1/\sqrt[3]{x} & \text{for } 1/\sqrt[3]{x} \leqq p \ \text{ or } \ x \geqq 1/p^3 \\ p & \text{for } 1/\sqrt[3]{x} > p \ \text{ or } \ x < 1/p^3 \end{cases}$$

Then

$$\int_0^8 \frac{dx}{\sqrt[3]{x}} = \lim_{p \to \infty} \int_0^8 [f(x)]_p \, dx$$

$$= \lim_{p \to \infty} \left[\int_0^{1/p^3} p \, dx + \int_{1/p^3}^8 \frac{dx}{\sqrt[3]{x}} \right]$$

$$= \lim_{p \to \infty} \left[\left. px \right|_0^{1/p^3} + \frac{3}{2} x^{2/3} \Big|_{1/p^3}^8 \right]$$

$$= \lim_{p \to \infty} \left[\frac{1}{p^2} + 6 - \frac{3}{2p^2} \right]$$

$$= 6$$

Thus the integral exists and has the value 6. Note that the above Lebesgue integrals can be evaluated as for Riemann integrals because of Theorem 4-17, page 57.

Note also that the Riemann integral, $(\mathcal{R}) \displaystyle\int_0^8 \dfrac{dx}{\sqrt[3]{x}}$, exists as an *improper integral* defined as

$$\lim_{\epsilon \to 0} (\mathcal{R}) \int_\epsilon^8 \frac{dx}{\sqrt[3]{x}} = \lim_{\epsilon \to 0} \left[\frac{3}{2} x^{2/3} \Big|_\epsilon^8 \right] = \lim_{\epsilon \to 0} \left[6 - \frac{3}{2} \epsilon^{2/3} \right] = 6, \text{ i.e. it has the same value as the Lebesgue}$$

integral above [see Appendix A, page 158]. The Lebesgue integral given above is not an "improper integral", however, even though the integrand is unbounded.

5.5. Graph the functions $f(x) = 1/\sqrt[3]{x}$ and $[f(x)]_p$, $0 < x \leqq 8$ of Problem 5.4 and thus illustrate geometrically the results of that problem.

The graphs of $f(x)$ and $[f(x)]_p$ are indicated in Fig. 5-1 and 5-2 respectively. The function $[f(x)]_p$ is often called a *truncated function* since $[f(x)]_p = p$ if $f(x) > p$.

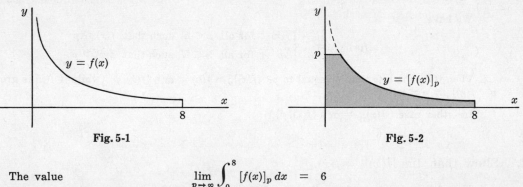

Fig. 5-1   Fig. 5-2

The value

$$\lim_{p \to \infty} \int_0^8 [f(x)]_p \, dx = 6$$

represents the limit of the area shown shaded in Fig. 5-2 as $p \to \infty$.

DEFINITION OF THE LEBESGUE INTEGRAL FOR ARBITRARY FUNCTIONS

5.6. Show that the functions $f^+(x)$ and $f^-(x)$, page 72, can be defined by $f^+(x) = \max \{f(x), 0\}$, $f^-(x) = \max \{0, -f(x)\}$.

We have

$$f^+(x) = \begin{cases} f(x) & \text{for all } x \in E \text{ such that } f(x) \geqq 0 \\ 0 & \text{for all } x \in E \text{ such that } f(x) < 0 \end{cases}$$

Then if $f(x) \geqq 0$, we have $f^+(x) = f(x) = \max\{f(x), 0\}$ while if $f(x) < 0$, $f^+(x) = 0 = \max\{f(x), 0\}$. In either case $f^+(x) = \max\{f(x), 0\}$.

By similar reasoning we show that $f^-(x) = \max\{0, -f(x)\}$.

5.7. Prove that (a) $f(x) = f^+(x) - f^-(x)$, (b) $|f(x)| = f^+(x) + f^-(x)$.

(a) We have $f^+(x) = f(x)$ if $f(x) \geqq 0$ and $f^-(x) = 0$ if $f(x) \geqq 0$. Then $f(x) = f^+(x) - f^-(x)$ if $f(x) \geqq 0$.

Similarly we have $f^+(x) = 0$ if $f(x) < 0$ and $f^-(x) = -f(x)$ if $f(x) < 0$. Then $f(x) = f^+(x) - f^-(x)$ if $f(x) < 0$.

Thus in all cases $f(x) = f^+(x) - f^-(x)$.

(b) If $f(x) \geqq 0$, then $|f(x)| = f(x) = f^+(x) + f^-(x)$. If $f(x) < 0$, then $|f(x)| = -f(x) = f^+(x) + f^-(x)$. Thus in all cases $|f(x)| = f^+(x) + f^-(x)$.

THEOREMS ON LEBESGUE INTEGRALS FOR UNBOUNDED FUNCTIONS

5.8. Prove Theorem 5-1, page 73: If $f(x) \geqq 0$, then $\displaystyle\int_E f(x)\,dx$ exists if and only if $\displaystyle\int_E [f(x)]_p\,dx$ is uniformly bounded.

From the definition of $[f(x)]_p$ it is seen that

$$[f(x)]_1 \leqq [f(x)]_2 \leqq \cdots \tag{1}$$

so that

$$\int_E [f(x)]_1\,dx \leqq \int_E [f(x)]_2\,dx \leqq \cdots \tag{2}$$

Thus the sequence whose pth term is

$$\int_E [f(x)]_p\,dx \tag{3}$$

is a monotonic increasing sequence. It follows that if this sequence is uniformly bounded, i.e. if there is a constant M such that for all p

$$\int_E [f(x)]_p\,dx < M$$

then

$$\lim_{p \to \infty} \int_E [f(x)]_p\,dx = \int_E f(x)\,dx$$

exists.

Conversely if $\displaystyle\int_E f(x)\,dx$ exists, then since $[f(x)]_p \leqq f(x)$ we have

$$\int_E [f(x)]_p\,dx \leqq \int_E f(x)\,dx < M$$

so that $\displaystyle\int_E [f(x)]_p\,dx$ is uniformly bounded [see Problem 5.41].

5.9. Prove Theorem 5-2, page 73, for the case where (a) $f(x) \geqq 0$, (b) $f(x)$ has arbitrary sign.

(a) If $0 \leqq f(x) \leqq g(x)$, then $[f(x)]_p \leqq [g(x)]_p$. Thus

$$\int_E [f(x)]_p\,dx \leqq \int_E [g(x)]_p\,dx \tag{1}$$

Then taking the limit as $p \to \infty$, we obtain

$$\int_E f(x)\, dx \;\leqq\; \int_E g(x)\, dx \tag{2}$$

which follows since $\int_E g(x)\, dx$ exists and since the integral on the left of (2) exists as a result of Problem 5.8.

(b) This follows at once from part (a) on replacing $f(x)$ by $|f(x)|$ which is non-negative.

5.10. Prove Theorem 5-4, page 73, for the case where $f(x) \geqq 0$.

Let $S = E[f(x) = \infty]$, i.e. the set of all x for which $f(x)$ is infinite. Then using the definition of $[f(x)]_p$, we have

$$\int_E [f(x)]_p\, dx \;\geqq\; \int_S [f(x)]_p\, dx \;=\; \int_S p\, dx \;=\; p\, m(S) \tag{1}$$

Taking the limit as $p \to \infty$, we have

$$\int_E f(x)\, dx \;\geqq\; \lim_{p \to \infty} p\, m(S) \tag{2}$$

Now if $m(S) > 0$, (2) will yield a contradiction since by hypothesis $\int_E f(x)\, dx$ is finite. Thus we must have $m(S) = 0$, i.e. $f(x)$ is finite almost everywhere.

For a proof of the theorem where $f(x)$ has arbitrary sign see Problem 5.47.

5.11. Prove Theorem 5-6, page 73, for the case where $f(x) \geqq 0$.

We have
$$\int_A [f(x)]_p\, dx \;\leqq\; \int_E [f(x)]_p\, dx \tag{1}$$

Then taking the limit as $p \to \infty$, it follows that

$$\int_A f(x)\, dx \;\leqq\; \int_E f(x)\, dx \tag{2}$$

i.e. the integral on the left of (2) exists since the integral on the right of (2) exists.

5.12. Prove Theorem 5-7, page 73, if $f(x) \geqq 0$.

We shall prove the theorem for any finite union of disjoint sets, i.e. where $E = E_1 \cup E_2 \cup \cdots \cup E_n$ and $E_1, E_2, \ldots, E_n$ are mutually disjoint. Since the theorem is true for bounded functions [see Problem 4.14, page 62] we have

$$\int_E [f(x)]_p\, dx \;=\; \int_{E_1} [f(x)]_p\, dx \;+\; \cdots \;+\; \int_{E_n} [f(x)]_p\, dx$$

Then taking the limit as $p \to \infty$, we have as required

$$\int_E f(x)\, dx \;=\; \int_{E_1} f(x)\, dx \;+\; \cdots \;+\; \int_{E_n} f(x)\, dx$$

since $f(x)$ is supposed to be Lebesgue integrable on E and thus on the measurable subsets $E_1, E_2, \ldots, E_n$.

5.13. Prove Theorem 5-7, page 73, if $f(x)$ has arbitrary sign.

The result of Problem 5.12 is true for $f^+(x)$ and $f^-(x)$, so that we have

$$\int_E f^+(x)\, dx \;=\; \int_{E_1} f^+(x)\, dx \;+\; \cdots \;+\; \int_{E_n} f^+(x)\, dx$$

$$\int_E f^-(x)\, dx \;=\; \int_{E_1} f^-(x)\, dx \;+\; \cdots \;+\; \int_{E_n} f^-(x)\, dx$$

Then by subtracting and noting that $\int_{E_k} f(x)\,dx = \int_{E_k} f^+(x)\,dx - \int_{E_k} f^-(x)\,dx$, we obtain the required result

$$\int_E f(x)\,dx \;=\; \int_{E_1} f(x)\,dx + \cdots + \int_{E_n} f(x)\,dx$$

5.14. Prove Theorem 5-8, page 74 for the case where $f(x) \geqq 0$.

We have by Problem 4.15, page 62,

$$\int_E [f(x)]_p\,dx \;=\; \int_{E_1} [f(x)]_p\,dx + \int_{E_2} [f(x)]_p\,dx + \cdots$$

$$\geqq \int_{E_1} [f(x)]_p\,dx + \cdots + \int_{E_n} [f(x)]_p\,dx$$

Then by letting $p \to \infty$, we have

$$\int_E f(x)\,dx \;\geqq\; \int_{E_1} f(x)\,dx + \cdots + \int_{E_n} f(x)\,dx \qquad (1)$$

Letting $n \to \infty$ in (1), we find

$$\int_E f(x)\,dx \;\geqq\; \int_{E_1} f(x)\,dx + \int_{E_2} f(x)\,dx + \cdots \qquad (2)$$

Now for $k = 1, 2, \ldots$ we also have

$$\int_{E_k} [f(x)]_p\,dx \;\leqq\; \int_{E_k} f(x)\,dx$$

so that on summing over k,

$$\int_{E_1} [f(x)]_p\,dx + \int_{E_2} [f(x)]_p\,dx + \cdots \;\leqq\; \int_{E_1} f(x)\,dx + \int_{E_2} f(x)\,dx + \cdots$$

i.e. $\qquad\qquad \int_E [f(x)]_p\,dx \;\leqq\; \int_{E_1} f(x)\,dx + \int_{E_2} f(x)\,dx + \cdots$

Letting $p \to \infty$, we thus obtain

$$\int_E f(x)\,dx \;\leqq\; \int_{E_1} f(x)\,dx + \int_{E_2} f(x)\,dx + \cdots \qquad (3)$$

From (2) and (3) it follows that

$$\int_E f(x)\,dx \;=\; \int_{E_1} f(x)\,dx + \int_{E_2} f(x)\,dx + \cdots$$

5.15. Prove Theorem 5-8, page 74, for the case where $f(x)$ has arbitrary sign.

From Problem 5.14 we have

$$\int_E f^+(x)\,dx \;=\; \int_{E_1} f^+(x)\,dx + \int_{E_2} f^+(x)\,dx + \cdots$$

$$\int_E f^-(x)\,dx \;=\; \int_{E_1} f^-(x)\,dx + \int_{E_2} f^-(x)\,dx + \cdots$$

Then by subtracting and using the fact that $\int_{E_k} f(x)\,dx = \int_{E_k} f^+(x)\,dx - \int_{E_k} f^-(x)\,dx$, we obtain the required result

$$\int_E f(x)\,dx \;=\; \int_{E_1} f(x)\,dx + \int_{E_2} f(x)\,dx + \cdots$$

5.16. Give an example to show that Theorem 5-8, page 74, need not be true if no mention is made of the integrability of $f(x)$ on E.

Let
$$f(x) = \begin{cases} n, & \dfrac{2n}{4n^2-1} < x \leqq \dfrac{1}{2n-1} \\ -n, & \dfrac{1}{2n+1} < x \leqq \dfrac{2n}{4n^2-1} \end{cases} \quad \text{for} \quad n = 1, 2, \ldots$$

Then for each n we have

$$\int_{1/(2n+1)}^{1/(2n-1)} f(x)\,dx = \int_{2n/(4n^2-1)}^{1/(2n-1)} n\,dx + \int_{1/(2n+1)}^{2n/(4n^2-1)} -n\,dx$$

$$= n\left(\frac{1}{2n-1} - \frac{2n}{4n^2-1}\right) - n\left(\frac{2n}{4n^2-1} - \frac{1}{2n+1}\right)$$

$$= 0$$

Now

$$\int_0^1 |f(x)|\,dx = \sum_{n=1}^{\infty} \int_{1/(2n+1)}^{1/(2n-1)} |f(x)|\,dx$$

$$= \sum_{n=1}^{\infty} \int_{1/(2n+1)}^{1/(2n-1)} n\,dx$$

$$= \sum_{n=1}^{\infty} \frac{n}{4n^2-1}$$

$$\geqq \sum_{n=1}^{\infty} \frac{n}{4n^2}$$

$$= \frac{1}{4}\sum_{n=1}^{\infty} \frac{1}{n}$$

But since the last series diverges, it follows that $\int_0^1 |f(x)|\,dx$ is infinite and so $\int_0^1 f(x)\,dx$ does not exist.

5.17. Prove Theorem 5-9, page 74, for the case where $f(x) \geqq 0$ and $g(x) \geqq 0$.

Let $h(x) = f(x) + g(x)$. Then if we let

$$[f(x)]_p = \begin{cases} f(x) & \text{for all } x \in E \text{ such that } f(x) \leqq p \\ p & \text{for all } x \in E \text{ such that } f(x) > p \end{cases}$$

with similar definitions for $[g(x)]_p$ and $[h(x)]_p$, we have

$$[h(x)]_p \leqq [f(x)]_p + [g(x)]_p \leqq [h(x)]_{2p}$$

Thus we have

$$\int_E [h(x)]_p\,dx \leqq \int_E [f(x)]_p\,dx + \int_E [g(x)]_p\,dx \leqq \int_E [h(x)]_{2p}\,dx$$

Letting $p \to \infty$, we find

$$\int_E h(x)\,dx \leqq \int_E f(x)\,dx + \int_E g(x)\,dx \leqq \int_E h(x)\,dx$$

which shows that
$$\int_E f(x)\,dx + \int_E g(x)\,dx = \int_E h(x)\,dx$$

i.e.
$$\int_E [f(x)+g(x)]\,dx = \int_E f(x)\,dx + \int_E g(x)\,dx$$

The result is easily extended to any finite number of functions.

5.18. Prove Theorem 5-9, page 74, for the case where $f(x)$ and $g(x)$ have arbitrary sign.

In case $f(x)$ and $g(x)$ have arbitrary sign and $h(x) = f(x) + g(x)$, the following six cases arise:

(1) $f(x) \geqq 0,\ g(x) \geqq 0,\ h(x) \geqq 0$ (4) $f(x) \geqq 0,\ g(x) < 0,\ h(x) < 0$

(2) $f(x) < 0,\ g(x) < 0,\ h(x) < 0$ (5) $f(x) < 0,\ g(x) \geqq 0,\ h(x) \geqq 0$

(3) $f(x) \geqq 0,\ g(x) < 0,\ h(x) \geqq 0$ (6) $f(x) < 0,\ g(x) \geqq 0,\ h(x) < 0$

Let $E_1, E_2, \ldots, E_6$ denote the respective sets in which each of these cases arises. It is then clear that $E = \bigcup\limits_{k=1}^{6} E_k$.

Now if we can show that

$$\int_{E_k} [f(x) + g(x)]\, dx \;=\; \int_{E_k} [f(x)]\, dx + \int_{E_k} [g(x)]\, dx \tag{1}$$

for $k = 1, 2, \ldots, 6$ the required result will follow by addition and use of Problem 5.15.

We have already established (1) for the case $k = 1$, i.e. the set E_1 in Problem 5.17. The result can also be established in a similar manner for any of the other values of k. As an example let us illustrate the case where $k = 4$. In such case we can write $h(x) = f(x) + g(x)$ as

$$[-g(x)] \;=\; [f(x)] + [-h(x)]$$

Then since $f(x) \geqq 0,\ -g(x) \geqq 0,\ -h(x) \geqq 0$ on the set E_4, we obtain as a consequence of Problem 5.17,

$$\int_{E_4} -g(x)\, dx \;=\; \int_{E_4} f(x)\, dx + \int_{E_4} -h(x)\, dx$$

Using Theorem 5-10 this becomes

$$-\int_{E_4} g(x)\, dx \;=\; \int_{E_4} f(x)\, dx - \int_{E_4} h(x)\, dx$$

or

$$\int_{E_4} h(x)\, dx \;=\; \int_{E_4} f(x)\, dx + \int_{E_4} g(x)\, dx$$

i.e.

$$\int_{E_4} [f(x) + g(x)]\, dx \;=\; \int_{E_4} f(x)\, dx + \int_{E_4} g(x)\, dx$$

5.19. Prove Theorem 5-3, page 73: (a) $f(x)$ is Lebesgue integrable on E if and only if $|f(x)|$ is Lebesgue integrable on E and (b) $\left| \int_E f(x)\, dx \right| \leqq \int_E |f(x)|\, dx$.

(a) From Problem 5.18 we have

$$\int_E f(x)\, dx \;=\; \int_E f^+(x)\, dx - \int_E f^-(x)\, dx$$

$$\int_E |f(x)|\, dx \;=\; \int_E f^+(x)\, dx + \int_E f^-(x)\, dx$$

Then if $f(x)$ is Lebesgue integrable on E, so also is $|f(x)|$; and conversely if $|f(x)|$ is Lebesgue integrable on E, so also is $f(x)$.

(b) From part (a) we have

$$\left| \int_E f(x)\, dx \right| \;=\; \left| \int_E f^+(x)\, dx - \int_E f^-(x)\, dx \right|$$

$$\leqq\; \left| \int_E f^+(x)\, dx \right| + \left| \int_E f^-(x)\, dx \right|$$

$$=\; \int_E f^+(x)\, dx + \int_E f^-(x)\, dx$$

$$=\; \int_E |f(x)|\, dx$$

5.20. Prove Theorem 5-11, page 74.

Since $g(x)$ is bounded, we have $|g(x)| \leqq M$ so that

$$|f(x)\, g(x)| \ \leqq \ M\, |f(x)|$$

Then we have

$$\int_E |f(x)\, g(x)|\, dx \ \leqq \ M \int_E |f(x)|\, dx$$

and the required result follows since $f(x)$, and thus $|f(x)|$, is integrable on E.

5.21. Prove Theorem 5-14, page 74:   If $f(x)$ is integrable on E, then given $\epsilon > 0$ there exist a $\delta > 0$ and a set $A \subset E$ such that

$$\left| \int_A f(x)\, dx \right| \ < \ \delta$$

if $m(A) < \delta$.

From Problem 5.19 it follows that $|f(x)|$ is integrable on E. Thus given $\epsilon_1 > 0$ there exists a natural number p_0 such that

$$\int_E (|f(x)| - [|f(x)|]_{p_0})\, dx \ < \ \epsilon_1 \tag{1}$$

where the integrand in (1) is non-negative.

Now if A is a measurable subset of E, we have by Theorem 5-6,

$$\int_A (|f(x)| - [|f(x)|]_{p_0})\, dx \ \leqq \ \int_E (|f(x)| - [|f(x)|]_{p_0})\, dx \tag{2}$$

Then from (1) and (2) we obtain

$$\int_A |f(x)|\, dx \ < \ \epsilon_1 + \int_A [|f(x)|]_{p_0}\, dx \tag{3}$$

But from the fact that $[|f(x)|]_{p_0} \leqq p_0$ we have

$$\int_A [|f(x)|]_{p_0}\, dx \ \leqq \ p_0\, m(A)$$

so that (3) can be written

$$\int_A |f(x)|\, dx \ < \ \epsilon_1 + p_0\, m(A) \tag{4}$$

Now let $\epsilon_1 = \epsilon/2$ and choose $m(A) < \delta = \epsilon/2p_0$. Then we have from (4),

$$\left| \int_A f(x)\, dx \right| \ \leqq \ \int_A |f(x)|\, dx \ < \ \epsilon$$

if $m(A) < \delta$, as required.

LEBESGUE'S DOMINATED CONVERGENCE THEOREM

5.22. Prove Lebesgue's dominated convergence theorem [Theorem 5-16, page 74]:   Let $\langle f_n(x) \rangle$ be a sequence of functions measurable on E such that $\lim\limits_{n \to \infty} f_n(x) = f(x)$.   Then if there exists a function $M(x)$ integrable on E such that $|f_n(x)| \leqq M(x)$, we have

$$\lim_{n \to \infty} \int_E f_n(x)\, dx \ = \ \int_E f(x)\, dx$$

From the fact that $|f_n(x)| \leqq M(x)$, we have $|f(x)| \leqq M(x)$ so that $f(x)$ is integrable. Now since

$$\left| \int_E [f(x) - f_n(x)] \, dx \right| \;\leq\; \int_E |f(x) - f_n(x)| \, dx$$

the required result will follow if we can prove that

$$\lim_{n \to \infty} \int_E |f(x) - f_n(x)| \, dx \;=\; 0 \tag{1}$$

To do this we use the sets of Problem 4.22, page 66, and write

$$\int_E |f(x) - f_n(x)| \, dx \;=\; \int_{S_n} |f(x) - f_n(x)| \, dx \;+\; \int_{R_n} |f(x) - f_n(x)| \, dx \tag{2}$$

On S_n we have $|f(x) - f_n(x)| < \epsilon$. Also since $|f_n(x)| \leq M(x)$ and $|f(x)| \leq M(x)$, we have $|f(x) - f_n(x)| \leq |f(x)| + |f_n(x)| \leq 2M(x)$. Then on using the mean-value theorem, (2) becomes

$$\int_E |f(x) - f_n(x)| \, dx \;\leq\; \epsilon \, m(S_n) + 2 \int_{R_n} M(x) \, dx \tag{3}$$

As $n \to \infty$ the last integral goes to zero since the infinite series defined by

$$\int_E M(x) \, dx \;=\; \int_{E_1} M(x) \, dx \;+\; \int_{E_2} M(x) \, dx \;+\; \cdots$$

$$=\; \int_{E_1} M(x) \, dx \;+\; \cdots \;+\; \int_{E_n} M(x) \, dx \;+\; \int_{R_n} M(x) \, dx$$

converges. Thus since $\lim_{n \to \infty} m(S_n) = m(E)$, we find from (3)

$$\overline{\lim_{n \to \infty}} \int_E |f(x) - f_n(x)| \, dx \;\leq\; \epsilon \, m(E)$$

Then letting $\epsilon \to 0$ we obtain (1) as required.

FATOU'S THEOREM

5.23. Prove Fatou's theorem [Theorem 5-19, page 75]: Let $\langle f_n(x) \rangle$ be a sequence of non-negative measurable functions defined on E and suppose that $\lim_{n \to \infty} f_n(x) = f(x)$. Then

$$\varliminf_{n \to \infty} \int_E f_n(x) \, dx \;\geq\; \int_E f(x) \, dx$$

Define as usual

$$[f(x)]_p \;=\; \begin{cases} f(x) & \text{for all } x \in E \text{ such that } f(x) \leq p \\ p & \text{for all } x \in E \text{ such that } f(x) < p \end{cases}$$

$$[f_n(x)]_p \;=\; \begin{cases} f_n(x) & \text{for all } x \in E \text{ such that } f_n(x) \leq p \\ p & \text{for all } x \in E \text{ such that } f_n(x) > p \end{cases}$$

Since $\lim_{n \to \infty} f_n(x) = f(x)$, we have

$$\lim_{n \to \infty} [f_n(x)]_p \;=\; [f(x)]_p$$

Thus by Lebesgue's theorem of bounded convergence,

$$\lim_{n \to \infty} \int_E [f_n(x)]_p \, dx \;=\; \int_E [f(x)]_p \, dx \tag{1}$$

Now since $[f_n(x)]_p \leq f_n(x)$ we have

$$\int_E f_n(x) \, dx \;\geq\; \int_E [f_n(x)]_p \, dx \tag{2}$$

Then taking $\varliminf_{n \to \infty}$ of both sides of (2), we have

$$\varliminf_{n \to \infty} \int_E f_n(x) \, dx \;\geq\; \varliminf_{n \to \infty} \int_E [f_n(x)]_p \, dx \tag{3}$$

But since the limit as $n \to \infty$ of the right hand side of (3) exists as in (1), we have

$$\lim_{n \to \infty} \int_E f_n(x)\, dx \;\geqq\; \int_E [f(x)]_p\, dx \tag{4}$$

Taking the limit as $p \to \infty$ in (4), we thus have as required

$$\lim_{n \to \infty} \int_E f_n(x)\, dx \;\geqq\; \int_E f(x)\, dx$$

5.24. Prove Theorem 5-20, page 75.

Suppose that $\displaystyle\lim_{n \to \infty} \int_E f_n(x)\, dx$ is finite. Then by Fatou's theorem we have

$$\int_E f(x)\, dx \;\leqq\; \lim_{n \to \infty} \int_E f_n(x)\, dx \tag{1}$$

so that the left side of (1) exists. Then since $f_n(x) \leqq f(x)$, we have

$$\int_E f_n(x)\, dx \;\leqq\; \int_E f(x)\, dx \tag{2}$$

so that on taking the limit as $n \to \infty$ in (2) we have

$$\lim_{n \to \infty} \int_E f_n(x)\, dx \;\leqq\; \int_E f(x)\, dx \tag{3}$$

Thus from (1) and (3) we have

$$\lim_{n \to \infty} \int_E f_n(x)\, dx \;=\; \int_E f(x)\, dx$$

Similarly if $\displaystyle\int_E f(x)\, dx$ is finite then since $f_n(x) \leqq f(x)$ we have by Lebesgue's dominated convergence theorem,

$$\lim_{n \to \infty} \int_E f_n(x)\, dx \;=\; \int_E \lim_{n \to \infty} f_n(x)\, dx \;=\; \int_E f(x)\, dx$$

Thus the required result is proved.

APPROXIMATION OF INTEGRABLE FUNCTIONS BY CONTINUOUS FUNCTIONS

5.25. Let $I_1 = [a_1, b_1]$ be an interval contained in $I = [a, b]$. Define $f(x) = \begin{cases} 1, & x \in I_1 \\ 0, & x \notin I_1 \end{cases}$.

Prove that given $\epsilon > 0$ there is a continuous function $g(x)$ defined on I such that

$$\int_a^b |f(x) - g(x)|\, dx \;<\; \epsilon$$

Choose points $P_1 : a_1 - \delta$ and $Q_1 : b + \delta$ in the intervals (a, a_1) and (b_1, b) respectively where $0 < \delta < \epsilon/2$ as indicated in Fig. 5-3. Let A_1 and B_1 denote the points where $x = a_1$, $f(x) = 1$ and $x = b_1$, $f(x) = 1$ respectively. Construct lines $P_1 A_1$ and $B_1 Q_1$.

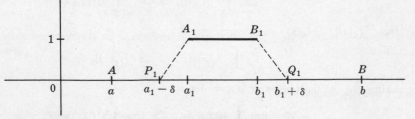

Fig. 5-3

Let $g(x)$ be the function whose graph is represented by $A P_1 A_1 B_1 Q_1 B$ in Fig. 5-3, i.e.

$$g(x) \;=\; \begin{cases} 0 & a \leqq x \leqq a_1 - \delta \\ 1 + (x - a_1)/\delta & a_1 - \delta \leqq x \leqq a_1 \\ 1 & a_1 \leqq x \leqq b_1 \\ 1 - (x - b_1)/\delta & b_1 \leqq x \leqq b_1 + \delta \\ 0 & b_1 + \delta \leqq x \leqq b \end{cases}$$

It is clear that $g(x)$ is continuous in $[a, b]$. Furthermore we have

$$\int_a^b |f(x) - g(x)|\, dx \;=\; \int_{a_1-\delta}^{a_1} \left(1 + \frac{x - a_1}{\delta}\right) dx + \int_{b_1}^{b_1+\delta} \left(1 - \frac{x - b_1}{\delta}\right) dx \;=\; 2\delta \;<\; \epsilon$$

and the required result follows.

5.26. Let $I_k = [a_k, b_k]$, $k = 1, 2, \ldots, n$, be n disjoint intervals contained in $I = [a, b]$. Let $f(x) = \begin{cases} 1, & x \in \cup I_k \\ 0, & x \notin \cup I_k \end{cases}$. Prove that given $\epsilon > 0$ there is a continuous function $g(x)$ defined on I such that

$$\int_a^b |f(x) - g(x)|\, dx \;<\; \epsilon$$

Let $f_k(x) = \begin{cases} 1, & x \in I_k \\ 0, & x \notin I_k \end{cases}$. Then by Problem 5.25 there is a continuous function $g_k(x)$ such that for $\epsilon > 0$

$$\int_a^b |f_k(x) - g_k(x)|\, dx \;<\; \frac{\epsilon}{n}, \qquad k = 1, \ldots, n$$

Thus letting $g(x) = \sum_{k=1}^n g_k(x)$, we have

$$\int_a^b |f(x) - g(x)|\, dx \;=\; \int_a^b \left| f(x) - \sum_{k=1}^n g_k(x) \right| dx$$

$$\leqq\; \sum_{k=1}^n \int_a^b |f_k(x) - g_k(x)|\, dx$$

$$<\; \epsilon$$

5.27. Let E be a measurable set contained in $I = [a, b]$ and suppose $f(x) = \begin{cases} 1, & x \in E \\ 0, & x \notin E \end{cases}$. Prove that given $\epsilon > 0$ there is a continuous function $g(x)$ such that

$$\int_a^b |f(x) - g(x)|\, dx \;<\; \epsilon$$

Since E is measurable, there is an open set $O \supset E$ such that if $\epsilon_1 > 0$,

$$m(E) \;\leqq\; m(O) \;<\; m(E) + \epsilon_1 \tag{1}$$

Now by Theorem 1-19, page 7, the set O can be expressed as a countable union of disjoint intervals I_k so that

$$m(O) \;=\; \sum_{k=1}^\infty m(I_k) \tag{2}$$

From (1) and (2) we see that

$$m(E) \;\leqq\; \sum_{k=1}^{n_0} m(I_k) \;+\; \sum_{k=n_0+1}^\infty m(I_k) \;<\; m(E) + \epsilon_1$$

Since the series converges we can choose n_0 sufficiently large so that

$$\sum_{k=n_0+1}^{\infty} m(I_k) \; < \; \epsilon_1 \tag{3}$$

and

$$m(E) \; \leqq \; \sum_{k=1}^{n_0} m(I_k) \; < \; m(E) + \epsilon_1 \tag{4}$$

Now consider the intervals $I_1, I_2, \ldots, I_{n_0}$ and let

$$f_{n_0}(x) \; = \; \begin{cases} 1, & x \in \bigcup_{k=1}^{n_0} I_k \\ 0, & x \notin \bigcup_{k=1}^{n_0} I_k \end{cases}$$

Then by Problem 5.26 there is a continuous function $g(x)$ such that

$$\int_a^b |f_{n_0}(x) - g(x)| \, dx \; < \; \epsilon_1 \tag{5}$$

Let $E_k = E \cap I_k$, $k = 1, 2, \ldots$, so that $E = \bigcup_{k=1}^{\infty} E_k$. Now if $k = 1, 2, \ldots, n_0$ we have since $I_k = (I_k - E_k) \cup E_k$,

$$\int_{I_k} |f(x) - f_{n_0}(x)| \, dx \; = \; \int_{I_k - E_k} |f(x) - f_{n_0}(x)| \, dx \; + \; \int_{E_k} |f(x) - f_{n_0}(x)| \, dx$$

$$= \; \int_{I_k - E_k} |0 - 1| \, dx \; + \; \int_{E_k} |1 - 1| \, dx$$

$$= \; \int_{I_k - E_k} dx \; = \; m(I_k - E_k) \; = \; m(I_k) - m(E_k)$$

If $k = n_0 + 1$, $n_0 + 2$, $\ldots$, then since $f_{n_0}(x) = 0$ for $x \in \bigcup_{k=n_0+1}^{\infty} I_k$,

$$\int_{I_k} |f(x) - f_{n_0}(x)| \, dx \; = \; \int_{I_k} |f(x)| \, dx \; \leqq \; \int_{I_k} dx \; = \; m(I_k)$$

Thus we have on using (3) and (4),

$$\int_a^b |f(x) - f_{n_0}(x)| \, dx \; = \; \sum_{k=1}^{n_0} \int_{I_k} |f(x) - f_{n_0}(x)| \, dx \; + \; \sum_{k=n_0+1}^{\infty} \int_{I_k} |f(x) - f_{n_0}(x)| \, dx$$

$$\leqq \; \sum_{k=1}^{n_0} [m(I_k) - m(E_k)] \; + \; \sum_{k=n_0+1}^{\infty} m(I_k)$$

$$= \; \left[\sum_{k=1}^{n_0} m(I_k) - m(E) \right] + \left[m(E) - \sum_{k=1}^{n_0} m(E_k) \right] + \sum_{k=n_0+1}^{\infty} m(I_k)$$

$$= \; \left[\sum_{k=1}^{n_0} m(I_k) - m(E) \right] + \sum_{k=n_0+1}^{\infty} m(E_k) + \sum_{k=n_0+1}^{\infty} m(I_k)$$

$$\leqq \; \sum_{k=1}^{n_0} m(I_k) - m(E) \; + \; \sum_{k=n_0+1}^{\infty} m(I_k) \; + \; \sum_{k=n_0+1}^{\infty} m(I_k)$$

$$\leqq \; \epsilon_1 + \epsilon_1 + \epsilon_1$$

$$= \; 3\epsilon_1$$

Thus on using (5) we find,

$$\int_a^b |f(x) - g(x)| \, dx \; \leqq \; \int_a^b |f(x) - f_{n_0}(x)| \, dx \; + \; \int_a^b |f_{n_0}(x) - g(x)| \, dx$$

$$< \; 3\epsilon_1 + \epsilon_1$$

$$= \; 4\epsilon_1$$

Choosing $\epsilon_1 = \epsilon/4$, the required result follows.

5.28. Let E_k, $k = 1, 2, \ldots, n$, be n disjoint measurable sets contained in $I = [a, b]$. Define $f_k(x) = \begin{cases} 1, & x \in E_k \\ 0, & x \notin E_k \end{cases}$ and $f(x) = \sum_{k=1}^{n} \alpha_k f_k(x)$ where α_k are constants. Prove that given $\epsilon > 0$, there is a continuous function $g(x)$ defined on I such that

$$\int_a^b |f(x) - g(x)| \, dx \; < \; \epsilon$$

By Problem 5.27 given $\epsilon_1 > 0$ we can find a continuous function $g_k(x)$ such that

$$\int_a^b |f_k(x) - g_k(x)| \, dx \; < \; \epsilon_1, \qquad k = 1, 2, \ldots, n$$

Define
$$g(x) \; = \; \sum_{k=1}^{n} \alpha_k \, g_k(x)$$

Then
$$\int_a^b |f(x) - g(x)| \, dx \; = \; \int_a^b \left| \sum_{k=1}^{n} \alpha_k [f_k(x) - g_k(x)] \right| dx$$
$$\leq \; \sum_{k=1}^{n} |\alpha_k| \int_a^b |f_k(x) - g_k(x)| \, dx$$
$$< \; \epsilon_1 \sum_{k=1}^{n} |\alpha_k|$$

Letting $\epsilon_1 \; = \; \epsilon \Big/ \sum_{k=1}^{n} |\alpha_k|$ we have as required,

$$\int_a^b |f(x) - g(x)| \, dx \; < \; \epsilon$$

5.29. Let $f(x)$ be bounded and measurable and thus integrable in $I = [a, b]$. Prove that given $\epsilon > 0$ there is a continuous function $g(x)$ defined on I such that

$$\int_a^b |f(x) - g(x)| \, dx \; < \; \epsilon$$

Choose subdivision points

$$\alpha \; = \; y_0 \; < \; y_1 \; < \; \cdots \; < \; y_{n-1} \; < \; y_n \; = \; \beta$$

Suppose that the largest of the values $y_k - y_{k-1}$ is denoted by δ and let

$$E_k \; = \; \{x : y_{k-1} \leq f(x) < y_k\}, \qquad k = 1, 2, \ldots, n$$

These sets are measurable since $f(x)$ is measurable.

Let $h_k(x) = \begin{cases} 1, & x \in E_k \\ 0, & x \in E_k \end{cases}$ and define

$$h(x) \; = \; \sum_{k=1}^{n} y_{k-1} \, h_k(x)$$

so that
$$h(x) \; = \; y_{k-1} \quad \text{for} \quad x \in E_k$$

Now by Problem 5.28, given $\epsilon_1 > 0$, there is a continuous function $g(x)$ such that

$$\int_a^b |h(x) - g(x)| \, dx \; < \; \epsilon_1$$

Letting δ be the largest of the quantities $\Delta y_k = y_k - y_{k-1}$, we find

$$\int_a^b |f(x) - g(x)|\, dx \;\leqq\; \int_a^b |f(x) - h(x)|\, dx \;+\; \int_a^b |h(x) - g(x)|\, dx$$

$$< \;\sum_{k=1}^{n} \int_{E_k} |f(x) - h(x)|\, dx \;+\; \epsilon_1$$

$$\leqq \;\sum_{k=1}^{n} \int_{E_k} \delta\, dx \;+\; \epsilon_1$$

$$\leqq \;\delta(b - a) + \epsilon_1 \;<\; \epsilon$$

if we choose $\epsilon_1 = \epsilon/2$ and $\delta < \epsilon/2(b-a)$. Thus the required result is proved.

5.30. Let $f(x)$ be unbounded and integrable in $[a, b]$. Prove that given $\epsilon > 0$, there is a continuous function $g(x)$ such that

$$\int_a^b |f(x) - g(x)|\, dx \;<\; \epsilon$$

Let

$$f_1(x) \;=\; \begin{cases} f(x), & f(x) \geqq 0 \\ 0, & f(x) \leqq 0 \end{cases}, \qquad f_2(x) \;=\; \begin{cases} 0, & f(x) \geqq 0 \\ -f(x), & f(x) \leqq 0 \end{cases}$$

so that $f(x) = f_1(x) - f_2(x)$ and

$$\int_a^b f(x)\, dx \;=\; \int_a^b f_1(x)\, dx \;-\; \int_a^b f_2(x)\, dx$$

Define

$$F_1(x) \;=\; \begin{cases} f_1(x), & f_1(x) \leqq n \\ n, & f_1(x) > n \end{cases}, \qquad F_2(x) \;=\; \begin{cases} f_2(x), & f_2(x) \leqq n \\ n, & f_2(x) > n \end{cases}$$

Then $F_1(x)$ and $F_2(x)$ are bounded and measurable in $[a, b]$. Thus by Problem 5.29, given $\epsilon_1 > 0$, there are continuous functions $g_1(x)$ and $g_2(x)$ such that

$$\int_a^b |F_1(x) - g_1(x)|\, dx \;<\; \epsilon_1, \qquad \int_a^b |F_2(x) - g_2(x)|\, dx \;<\; \epsilon_1$$

Also we can choose n so large that

$$\int_a^b |f_1(x) - F_1(x)|\, dx \;<\; \epsilon_1, \qquad \int_a^b |f_2(x) - F_2(x)|\, dx \;<\; \epsilon_1$$

If we call $g(x) = g_1(x) - g_2(x)$, then

$$|f(x) - g(x)| \;=\; |f_1(x) - f_2(x) - g_1(x) + g_2(x)|$$
$$\leqq\; |f_1(x) - F_1(x) + F_1(x) - g_1(x)| + |f_2(x) - F_2(x) + F_2(x) - g_2(x)|$$
$$\leqq\; |f_1(x) - F_1(x)| + |F_1(x) - g_1(x)| + |f_2(x) - F_2(x)| + |F_2(x) - g_2(x)|$$

Thus

$$\int_a^b |f(x) - g(x)|\, dx \;\leqq\; \int_a^b |f_1(x) - F_1(x)|\, dx \;+\; \int_a^b |F_1(x) - g_1(x)|\, dx$$
$$+\; \int_a^b |f_2(x) - F_2(x)|\, dx \;+\; \int_a^b |F_2(x) - g_2(x)|\, dx$$
$$<\; \epsilon_1 + \epsilon_1 + \epsilon_1 + \epsilon_1 \;=\; 4\epsilon_1$$

and the required result follows on choosing $\epsilon_1 = \epsilon/4$.

5.31. Prove Theorem 5-23, page 76: If $f(x)$ is integrable [and not necessarily bounded] in $[a, b]$, then

$$\lim_{h \to 0} \int_a^b |f(x + h) - f(x)|\, dx \;=\; 0$$

Let the interval $[a, b]$ be enclosed in an interval $[c, d]$ as indicated in Fig. 5-4 below and assume that $f(x)$ is suitably extended so as to be integrable in $[c, a]$ and $[b, d]$.

By Problem 5.30 there exists a continuous function $g(x)$ in $[c,d]$ so that given $\epsilon_1 > 0$

$$\int_c^d |f(x) - g(x)|\, dx < \epsilon_1$$

Fig. 5-4

Also since $g(x)$ is continuous in $[c,d]$, it follows that for every $\epsilon_1 > 0$ there is a $\delta > 0$ such that for all x in $[a,b]$,

$$|g(x+h) - g(x)| < \frac{\epsilon_1}{b-a} \quad \text{whenever} \quad |h| < \delta$$

where we can suppose that $|h|$ is less than the minimum of $a-c$ and $d-b$. Thus

$$\int_a^b |g(x+h) - g(x)|\, dx \leq \int_a^b \frac{\epsilon_1}{b-a}\, dx = \epsilon_1$$

Now

$$f(x+h) - f(x) = [f(x+h) - g(x+h)] + [g(x+h) - g(x)] + [g(x) - f(x)]$$

from which we have

$$|f(x+h) - f(x)| \leq |f(x+h) - g(x+h)| + |g(x+h) - g(x)| + |f(x) - g(x)|$$

so that for $|h| < \delta$,

$$\int_a^b |f(x+h) - f(x)|\, dx \leq \int_a^b |f(x+h) - g(x+h)|\, dx + \int_a^b |g(x+h) - g(x)|\, dx$$
$$+ \int_a^b |f(x) - g(x)|\, dx$$
$$= \int_{a+h}^{b+h} |f(x) - g(x)|\, dx + \int_a^b |g(x+h) - g(x)|\, dx$$
$$+ \int_a^b |f(x) - g(x)|\, dx$$
$$\leq \int_c^d |f(x) - g(x)|\, dx + \int_a^b |g(x+h) - g(x)|\, dx$$
$$+ \int_c^d |f(x) - g(x)|\, dx$$
$$< \epsilon_1 + \epsilon_1 + \epsilon_1$$
$$= 3\epsilon_1$$

Choosing $\epsilon_1 = \epsilon/3$, we see that

$$\int_a^b |f(x+h) - f(x)|\, dx < \epsilon \quad \text{for} \quad |h| < \delta$$

i.e.

$$\lim_{h \to 0} \int_a^b |f(x+h) - f(x)|\, dx = 0$$

LEBESGUE INTEGRALS ON BOUNDED SETS

5.32. Prove that if $f(x)$ is measurable on an unbounded set E, then

$$\left| \int_E f(x)\, dx \right| \leq \int_E |f(x)|\, dx$$

By Problem 5.19 we have

$$\left| \int_{E \cap [a,b]} f(x)\, dx \right| \leq \int_{E \cap [a,b]} |f(x)|\, dx$$

Then taking the limit as $a \to -\infty$ and $b \to \infty$ the required result follows.

Supplementary Problems

DEFINITION OF THE LEBESGUE INTEGRAL FOR UNBOUNDED FUNCTIONS

5.33. Determine $[f(x)]_p$ if (a) $f(x) = 1/x$, (b) $f(x) = e^{1/x^2}$ for $x > 0$.

5.34. Prove that $\int_0^4 \dfrac{dx}{\sqrt{x}}$ exists as a Lebesgue integral and find its value.   *Ans.* 4

5.35. Prove that $\int_0^2 \dfrac{dx}{x^2}$ does not exist.

5.36. Investigate the existence of (a) $\int_2^{10} \dfrac{dx}{\sqrt[3]{x-2}}$, (b) $\int_4^5 \dfrac{dx}{(x-4)^{3/2}}$.

5.37. Prove that $f(x) = x^{-q}$ is Lebesgue integrable in $(0,1)$ or not according as $q < 1$ or not.

5.38. Show that if $f(x)$ is bounded, measurable and non-negative, then the definition (2) on page 72 for the Lebesgue integral reduces to that given in Chapter 4.

5.39. Work Problem 5.3, page 77 by direct use of the definition of a measurable function, i.e. without relying on Problem 3.18, page 48.

5.40. Prove that (a) $[f(x)]_1 \leqq [f(x)]_2 \leqq \cdots$, (b) $\int_E [f(x)]_1 \, dx \leqq \int_E [f(x)]_2 \, dx \leqq \cdots$. Thus verify equations (1) and (2) of Problem 5.8.

5.41. Prove that if $f(x) \geqq 0$ then (a) $[f(x)]_p \leqq f(x)$ and (b) $\int_E [f(x)]_p \, dx \leqq \int_E f(x) \, dx$ if $f(x)$ is integrable on E.

5.42. Find (a) $f^+(x)$ and (b) $f^-(x)$ for the function $f(x) = 3 + e^x - e^{-x}$, $-3 \leqq x \leqq 3$.

5.43. (a) Prove that if $f(x)$ is bounded and measurable, then so also are $f^+(x)$ and $f^-(x)$. (b) Use the result of part (a) to show that the definition of the Lebesgue integral given by (7), page 72, reduces to that given in Chapter 4 in case $f(x)$ is bounded and measurable.

5.44. Show that if $f(x)$ is unbounded and non-negative, then the definition of the Lebesgue integral given by (7), page 72, reduces to that given by (2), page 72.

5.45. Let E be the interval $(-8, 8)$ and $f(x) = (x^{1/3} - 1)/x^{2/3}$. Use the definition to determine whether or not $\int_E f(x) \, dx$ exists. If it exists find its value.

5.46. Give a geometric interpretation to the result of Problem 5.45.

THEOREMS INVOLVING LEBESGUE INTEGRALS OF UNBOUNDED FUNCTIONS

5.47. Prove Theorem 5-4, page 73, if $f(x)$ has arbitrary sign [compare Problem 5.10].

5.48. Prove Theorem 5-5, page 73, if (a) $f(x) \geqq 0$, (b) $f(x)$ has arbitrary sign.

5.49. Prove Theorem 5-6, page 73, if $f(x)$ has arbitrary sign [compare Problem 5.11].

5.50. Prove Theorem 5-9, page 74, for the cases (a) $f(x) < 0$, $g(x) < 0$, (b) $f(x) \geqq 0$, $g(x) < 0$, $h(x) = f(x) + g(x) \geqq 0$, (c) $f(x) < 0$, $g(x) \geqq 0$, $h(x) = f(x) + g(x) \geqq 0$, (d) $f(x) < 0$, $g(x) \geqq 0$, $h(x) = f(x) + g(x) < 0$.

5.51. Generalize Theorem 5-9 to any finite number of functions.

5.52. Prove Theorem 5-10, page 74.

5.53. Prove Theorem 5-12, page 74, for the cases (a) $f(x) \geqq 0$, $g(x) \geqq 0$, (b) $f(x)$, $g(x)$ have arbitrary sign.

5.54. Prove Theorem 5-13, page 74.

5.55. If c_1, c_2 are any constants and $f_1(x), f_2(x)$ are integrable on E, prove that

$$\int_E [c_1 f_1(x) + c_2 f_2(x)]\, dx \;=\; c_1 \int_E f_1(x)\, dx \,+\, c_2 \int_E f_2(x)\, dx$$

5.56. Generalize Problem 5.55 to any finite number of functions.

5.57. If $\int_E |f(x) - g(x)|\, dx = 0$, prove that $f(x) = g(x)$ almost everywhere on E.

5.58. Prove Theorem 5-15, page 74.

LEBESGUE'S DOMINATED CONVERGENCE THEOREM

5.59. Discuss conditions under which Lebesgue's dominated convergence theorem reduces to Lebesgue's theorem on bounded convergence.

5.60. Let $\;f_n(x) = \begin{cases} n, & 1/n^3 \le x \le 8/n^3 \\ 0, & 0 \le x < 1/n^3 \text{ or } 8/n^3 < x \le 1 \end{cases}$

    (a) Prove that $\lim\limits_{n \to \infty} f_n(x) = 0$ in $[0,1]$ but that the convergence is not uniform.

    (b) Prove that $|f_n(x)| \le M(x)$ where $M(x) = \begin{cases} 2/\sqrt[3]{x}, & 0 < x \le 1 \\ 0, & x = 0 \end{cases}$.

    (c) Is it true that $\lim\limits_{n \to \infty} \int_0^1 f_n(x)\, dx = \int_0^1 \lim\limits_{n \to \infty} f_n(x)\, dx$? Explain.

5.61. Suppose that $\lim\limits_{n \to \infty} f_n(x) = f(x)$ almost everywhere in E and that $|f_n(x)| \le M(x)$ almost everywhere in E where $M(x)$ is integrable on E. If $v(x)$ is bounded and measurable on E, prove that

$$\lim_{n \to \infty} \int_E f_n(x)\, v(x)\, dx \;=\; \int_E f(x)\, v(x)\, dx$$

5.62. Prove Theorem 5-17, page 75.

5.63. Prove Theorem 5-18, page 75.

5.64. Are the results of Problem 5.61 and Theorem 5-18 still valid if $v(x)$ is integrable on E? Explain.

5.65. Work Problem A.60, page 173, if the integrals are Lebesgue integrals.

FATOU'S THEOREM

5.66. Prove that $\varliminf\limits_{n \to \infty} \int_E f_n(x)\, dx \geqq \int_E \varliminf\limits_{n \to \infty} f_n(x)\, dx$ where $\langle f_n(x) \rangle$ is a sequence of non-negative functions defined on E. Discuss the relationship with Fatou's theorem.

5.67. Prove that if the conditions of Problem 5.66 are satisfied, then

$$\varlimsup_{n \to \infty} \int_E f_n(x)\, dx \;\le\; \int_E \varlimsup_{n \to \infty} f_n(x)\, dx$$

Discuss the relationship with Fatou's theorem and with Problem 5.66.

5.68. Let

$$f_n(x) \;=\; \begin{cases} n^3, & 1/n^3 \leq x \leq 8/n^3 \\ 0, & 0 \leq x < 1/n^3 \ \text{ or } \ 8/n^3 < x \leq 1 \end{cases}$$

    (*a*) Is it true that $\displaystyle \lim_{n \to \infty} \int_0^1 f_n(x)\, dx \;=\; \int_0^1 \lim_{n \to \infty} f_n(x)\, dx$?

    (*b*) Is it true that $\displaystyle \varliminf_{n \to \infty} \int_0^1 f_n(x)\, dx \;\geq\; \int_0^1 \lim_{n \to \infty} f_n(x)\, dx$? Explain the relationship between parts
    (*a*) and (*b*).

LEBESGUE INTEGRALS ON UNBOUNDED SETS

5.69. Prove that $\displaystyle \int_1^\infty \frac{dx}{x^2}$ exists as a Lebesgue integral and find its value.   *Ans.* 1

5.70. Prove that $\displaystyle \int_0^\infty \frac{\sin^2 x}{x^2}\, dx$ exists as both a Lebesgue and a Riemann integral.

5.71. Prove Lebesgue's dominated convergence theorem for the case where the set E is unbounded.

5.72. Prove Fatou's theorem for the case where the set E is unbounded.

5.73. Is the result of Problem 5.61 still true if E is unbounded?   Explain.

5.74. Let $\displaystyle f_n(x) = \begin{cases} x/n^2, & 0 < x < n \\ 0, & \text{otherwise} \end{cases}$.   (*a*) Evaluate $\displaystyle \lim_{n \to \infty} \int_0^n f_n(x)\, dx$.   (*b*) Is the result in (*a*) the
same as $\displaystyle \int_0^\infty \lim_{n \to \infty} f_n(x)\, dx$?   Explain.

5.75. Can we define a Cauchy principal value for Lebesgue integrals?   Explain.

Chapter 6

Differentiation and Integration

INDEFINITE INTEGRALS

If $f(x)$ is integrable, we call

$$F(x) \;=\; \int_a^x f(u)\,du \tag{1}$$

an *indefinite integral* or briefly an *integral* of $f(x)$. Note that the dummy symbol u is used to avoid confusion with x in the upper limit of integration. If any constant is added to the right side of (1) the result is also called an indefinite integral.

SOME THEOREMS ON INDEFINITE RIEMANN INTEGRALS

If (1) is a Riemann integral, we have the following theorems [see Appendix A, page 157].

Theorem 6-1.   If $f(x)$ is bounded and Riemann integrable on $[a,b]$, then

$$F(x) \;=\; \int_a^x f(u)\,du$$

is continuous in $[a,b]$.

Theorem 6-2.   If $f(x)$ is bounded and Riemann integrable on $[a,b]$, then

$$F'(x) \;=\; \frac{d}{dx}F(x) \;=\; \frac{d}{dx}\int_a^x f(u)\,du \;=\; f(x) \tag{2}$$

at each point of continuity of $f(x)$.

Theorem 6-3 [**Fundamental theorem of calculus**].   Let $f(x)$ be Riemann integrable on $[a,b]$ and suppose that there exists a function $F(x)$ continuous on $[a,b]$ such that $f(x) = F'(x)$. Then

$$\int_a^b f(x)\,dx \;=\; F(b) - F(a) \tag{3}$$

or

$$\int_a^x f(u)\,du \;=\; F(x) - F(a) \tag{4}$$

It is the purpose of this chapter to investigate the relationship of differentiation and integration corresponding to the above theorems for the case of indefinite Lebesgue integrals.

MONOTONIC FUNCTIONS

A function $f(x)$ in $[a,b]$ is said to be *monotonic increasing* if $f(x_2) \geqq f(x_1)$ for $x_2 > x_1$, and *monotonic decreasing* if $f(x_2) \leqq f(x_1)$ for $x_2 > x_1$. If $f(x)$ is either monotonic increasing or monotonic decreasing, it is called *monotonic*.

The *jump* of $f(x)$ at x_0 is defined as $f(x_0 + 0) - f(x_0 - 0)$ where $f(x_0 + 0) = \lim_{h \to 0+} f(x_0 + h)$, $f(x_0 - 0) = \lim_{h \to 0-} f(x_0 + h)$. If $f(x)$ is monotonic increasing, the jump at any point is non-negative. If $f(x)$ is monotonic decreasing, the jump is non-positive. If $f(x)$ is continuous at x_0, then the jump at x_0 is equal to zero.

If $f(x)$ is monotonic increasing, then $-f(x)$ is monotonic decreasing and conversely. Because of this result many properties proved for monotonic increasing functions are also true for monotonic decreasing functions.

SOME THEOREMS ON MONOTONIC FUNCTIONS

Theorem 6-4. If $f(x)$ is monotonic increasing in $[a, b]$, then there can be at most a finite number of discontinuities with jump greater than some given $\epsilon > 0$. A similar result is true for monotonic decreasing functions.

Theorem 6-5. If $f(x)$ is monotonic increasing [or monotonic decreasing] in $[a, b]$, then the set of discontinuities of $f(x)$ can at most be denumerable.

FUNCTIONS OF BOUNDED VARIATION

Suppose that the interval $[a, b]$ is divided into n parts by use of the points of subdivision $a = x_0 < x_1 < \cdots < x_n = b$. Then we say that $f(x)$ is of *bounded variation* in $[a, b]$ if

$$\sum_{k=1}^{n} |f(x_k) - f(x_{k-1})| \tag{5}$$

is bounded for all possible partitions or modes of subdivision.

The least upper bound of all the sums (5) for all possible modes of subdivision is called the *total variation* of $f(x)$ in $[a, b]$ and will be denoted by $\mathcal{V}_a^b(f)$ or briefly $\mathcal{V}_a^b$.

SOME THEOREMS ON FUNCTIONS OF BOUNDED VARIATION

Theorem 6-6. A monotonic increasing [or monotonic decreasing] function in $[a, b]$ is of bounded variation in $[a, b]$.

Theorem 6-7. If $f(x)$ is of bounded variation in $[a, b]$, then it is bounded in $[a, b]$.

Theorem 6-8. If $f(x)$ and $g(x)$ are of bounded variation in $[a, b]$, so also are $f(x) + g(x)$, $f(x) - g(x)$, $f(x)g(x)$ and $f(x)/g(x)$ where in the last case $g(x) \neq 0$.

Theorem 6-9 [**Jordan decomposition theorem**]. A function $f(x)$ is of bounded variation if and only if it can be expressed as the difference $G(x) - H(x)$ of two bounded monotonic increasing functions $G(x)$ and $H(x)$.

Theorem 6-10. If $f(x)$ is of bounded variation in $[a, b]$, then it can have at most a denumerable set of discontinuities in $[a, b]$, and at every point c in (a, b) the right and left hand limits $\lim_{x \to c+} f(x) = f(c + 0)$ and $\lim_{x \to c-} f(x) = f(c - 0)$ exist.

Theorem 6-11. If $f(x)$ has a bounded derivative in $[a, b]$, then $f(x)$ is of bounded variation in $[a, b]$. However, if $f(x)$ is continuous in $[a, b]$ it need not be of bounded variation [see Problem 6.5].

DERIVATES OF A FUNCTION

The *derivative* of $f(x)$ is defined as

$$D f(x) = f'(x) = \lim_{h \to 0} \frac{f(x+h) - f(x)}{h} \tag{6}$$

and may or may not exist. Let us define four quantities

$$D^+ f(x) = \overline{\lim_{h \to 0+}} \frac{f(x+h) - f(x)}{h} \tag{7}$$

$$D^- f(x) = \overline{\lim_{h \to 0-}} \frac{f(x+h) - f(x)}{h} \tag{8}$$

$$D_+ f(x) = \lim_{h \to 0+} \frac{f(x+h) - f(x)}{h} \tag{9}$$

$$D_- f(x) = \lim_{h \to 0-} \frac{f(x+h) - f(x)}{h} \tag{10}$$

called *derivates* of $f(x)$, which are either finite, positively infinite or negatively infinite.

In case $\qquad\qquad D^+ f(x) = D_+ f(x)$

then $f(x)$ has a *right hand derivative* denoted by $f'_+(x)$.

In case $\qquad\qquad D^- f(x) = D_- f(x)$

then $f(x)$ has a *left hand derivative* denoted by $f'_-(x)$.

In case $\qquad D^+ f(x) = D_+ f(x) = D^- f(x) = D_- f(x)$

then $f(x)$ has a derivative $f'(x)$. Conversely if $f(x)$ has a derivative, then all of the four derivates are equal.

DERIVATIVES OF FUNCTIONS OF BOUNDED VARIATION

The following theorems are of great importance.

Theorem 6-12. If a function is monotonic, then it has a derivative almost everywhere.

Theorem 6-13. If a function is of bounded variation, then it has a derivative almost everywhere.

The following theorem is often useful.

Theorem 6-14. If $\sum_{k=1}^{\infty} f_k(x)$ is a series of functions of bounded variation which converges to $s(x)$ in $[a, b]$, then almost everywhere in $[a, b]$,

$$s'(x) = \sum_{k=1}^{\infty} f'_k(x)$$

ABSOLUTE CONTINUITY

A function $f(x)$ is said to be absolutely continuous in $[a, b]$ if given $\epsilon > 0$ there exists $\delta > 0$ such that

$$\sum_{k=1}^{n} |f(x_k + h_k) - f(x_k)| < \epsilon \tag{11}$$

for every finite set of intervals $(x_k, x_k + h_k)$ such that

$$\sum_{k=1}^{n} |h_k| < \delta \tag{12}$$

The definition is sometimes used with $x_k + h_k = x_{k+1}$, i.e. $h_k = x_{k+1} - x_k$.

SOME THEOREMS ON ABSOLUTE CONTINUITY

Theorem 6-15. An absolutely continuous function is continuous but the converse need not be true.

Theorem 6-16. If $f(x)$ and $g(x)$ are absolutely continuous in $[a, b]$, so also are $f(x) + g(x)$, $f(x) - g(x)$, $f(x) g(x)$ and $f(x)/g(x)$ where in the last case $g(x) \neq 0$.

Theorem 6-17. An absolutely continuous function is of bounded variation but the converse need not be true.

Theorem 6-18. An absolutely continuous function has a derivative almost everywhere.

Theorem 6-19. If $f(x)$ is absolutely continuous in $[a, b]$ and $f'(x) = 0$ almost everywhere in $[a, b]$, then $f(x)$ is a constant in $[a, b]$.

Theorem 6-20. Let $f(x)$ be absolutely continuous and strictly increasing in $[a, b]$ [i.e. $f(x_k) > f(x_{k-1})$ for $x_k > x_{k-1}$]. Then if $g(u)$ is absolutely continuous in $[f(a), f(b)]$, $g(f(x))$ is absolutely continuous in $[a, b]$.

In connection with Theorems 6-15 and 6-17 it is of interest to note that even a continuous function of bounded variation may not be absolutely continuous.

THEOREMS ON INDEFINITE LEBESGUE INTEGRALS

Theorem 6-21. If $f(x)$ is (Lebesgue) integrable on $[a, b]$, then

$$F(x) = \int_a^x f(u)\, du$$

is continuous and of bounded variation in $[a, b]$.

Theorem 6-22. If $g(x)$ is integrable on $[a, b]$ and

$$\int_a^x g(u)\, du = 0$$

for all x in $[a, b]$, then $g(x) = 0$ almost everywhere in $[a, b]$.

Theorem 6-23. If $g(x)$ is continuous and monotonic increasing in $[a, b]$, then $g'(x)$ is integrable on $[a, b]$ and

$$\int_a^b g'(x)\, dx \leq g(b) - g(a)$$

or on replacing b by x,

$$\int_a^x g'(u)\, du \leq g(x) - g(a)$$

Theorem 6-24. A function $F(x)$ is an indefinite integral if and only if it is absolutely continuous.

Theorem 6-25. If $F(x)$ is absolutely continuous, then

$$\int_a^b F'(x)\,dx \;=\; F(b) - F(a)$$

or replacing b by x,

$$\int_a^x F'(u)\,du \;=\; F(x) - F(a)$$

Theorem 6-26. If $F'(x)$ exists everywhere in $[a,b]$ and is bounded, then

$$\int_a^b F'(x)\,dx \;=\; F(b) - F(a)$$

or

$$\int_a^x F'(u)\,du \;=\; F(x) - F(a)$$

INTEGRATION BY PARTS

Theorem 6-27. Let $F(x)$ and $G(x)$ be indefinite integrals of the integrable functions $f(x)$ and $g(x)$ respectively. Then

$$\begin{aligned}
\int_a^b F(x)\,g(x)\,dx \;&=\; F(x)\,G(x)\Big|_a^b - \int_a^b f(x)\,G(x)\,dx \\
&=\; F(b)\,G(b) - F(a)\,G(a) - \int_a^b f(x)\,G(x)\,dx
\end{aligned}$$

A related theorem is the following.

Theorem 6-28. If $F(x)$ and $G(x)$ are absolutely continuous, then

$$\int_a^b F(x)\,G'(x)\,dx \;=\; F(x)\,G(x)\Big|_a^b - \int_a^b F'(x)\,G(x)\,dx$$

CHANGE OF VARIABLES

Theorem 6-29. If $f(x)$ is integrable in $[a,b]$ and $x = \Phi(u)$ is an indefinite integral of a non-negative integrable function $\phi(u)$, then

$$\int_a^b f(x)\,dx \;=\; \int_\alpha^\beta f(\Phi(u))\,\Phi'(u)\,du \;=\; \int_\alpha^\beta f(\Phi(u))\,\phi(u)\,du$$

where α and β are such that $\Phi(\alpha) = a,\;\; \Phi(\beta) = b$.

Solved Problems

MONOTONIC FUNCTIONS

6.1. Let $f(x)$ be monotonic increasing in $[a, b]$ and suppose that subdivision points are chosen such that $a < x_1 < \cdots < x_{n-1} < b$. Prove that

$$f(a + 0) - f(a) + \sum_{k=1}^{n-1} [f(x_{k+0}) - f(x_{k-0})] + f(b) - f(b - 0) \;\leqq\; f(b) - f(a)$$

Choose points $c_0, \ldots, c_{n-1}$ such that

$$a < c_0 < x_1 < c_1 < x_2 < \cdots < x_{n-1} < c_{n-1} < b$$

Then we have

$$f(a + 0) - f(a) \;\leqq\; f(c_0) - f(a)$$
$$f(x_1 + 0) - f(x_1 - 0) \;\leqq\; f(c_1) - f(c_0)$$
$$f(x_2 + 0) - f(x_2 - 0) \;\leqq\; f(c_2) - f(c_1)$$
$$\cdots\cdots\cdots\cdots\cdots\cdots\cdots\cdots\cdots\cdots\cdots\cdots\cdots\cdots\cdots$$
$$f(x_{n-1} + 0) - f(x_{n-1} - 0) \;\leqq\; f(c_{n-1}) - f(c_{n-2})$$
$$f(b) - f(b - 0) \;\leqq\; f(b) - f(c_{n-1})$$

Adding, we obtain the required result.

An analogous result can be proved for the case where $f(x)$ is monotonic decreasing [see Problem 6.32].

6.2. Prove Theorem 6-4, page 96: If $f(x)$ is monotonic increasing in $[a, b]$, then there can be at most a finite number of discontinuities with jump greater than some given $\epsilon > 0$.

Suppose that the number of discontinuities with jump greater than ϵ is p. Then by Problem 6.1 we have

$$p\epsilon \;\leqq\; f(b) - f(a)$$

from which we see that p must be finite.

For the case where $f(x)$ is monotonic decreasing, see Problem 6.33.

6.3. Prove Theorem 6-5, page 96: If $f(x)$ is monotonic increasing [or monotonic decreasing] in $[a, b]$, then the set of discontinuities of $f(x)$ can at most be denumerable.

Let A be the set of all discontinuities of $f(x)$ in $[a, b]$ and let A_k be the set of all discontinuities of $f(x)$ for which the jump is greater than $1/k$. Then we have

$$A = \bigcup_{k=1}^{\infty} A_k$$

But since each of the sets A_k can have at most a finite number of discontinuities by Problem 6.2, it follows that A can have at most a denumerable set of discontinuities.

For the proof in case $f(x)$ is monotonic decreasing see Problem 6.34.

FUNCTIONS OF BOUNDED VARIATION

6.4. Prove that (a) a monotonic increasing function and (b) a monotonic decreasing function in $[a, b]$ are of bounded variation.

(a) Suppose that a partition of $[a, b]$ is $a = x_0 < x_1 < \cdots < x_n = b$. Then if the function is monotonic increasing we have $f(x_k) \geqq f(x_{k-1})$. Thus

$$\sum_{k=1}^{n} |f(x_k) - f(x_{k-1})| = \sum_{k=1}^{n} [f(x_k) - f(x_{k-1})] = f(b) - f(a)$$

Then the sums

$$\sum_{k=1}^{n} |f(x_k) - f(x_{k-1})|$$

are bounded for all possible partitions of $[a, b]$ so that $f(x)$ is of bounded variation in $[a, b]$.

(b) In this case we have $f(x_k) \leq f(x_{k-1})$ so that

$$\sum_{k=1}^{n} |f(x_k) - f(x_{k-1})| = \sum_{k=1}^{n} [f(x_{k-1}) - f(x_k)] = f(a) - f(b)$$

and the required result follows as in part (a).

6.5. Give an example to show that a continuous function is not necessarily of bounded variation.

Consider $f(x) = \begin{cases} x \sin(1/x), & 0 < x \leq 2/\pi \\ 0, & x = 0 \end{cases}$ and choose as partition of $[0, 2/\pi]$ the points

$$0, \quad \frac{2}{(2n+1)\pi}, \quad \frac{2}{(2n)\pi}, \quad \frac{2}{(2n-1)\pi}, \quad \cdots, \quad \frac{2}{2\pi}, \quad \frac{2}{\pi}$$

which we can denote respectively by $x_0, x_1, \ldots, x_{2n+1}$. Then

$$\sum_{k=1}^{2n} |f(x_k) - f(x_{k-1})| = \frac{2}{\pi} + \frac{2}{2\pi} + \frac{2}{3\pi} + \cdots + \frac{2}{(2n+1)\pi}$$

$$= \frac{2}{\pi}\left[1 + \frac{1}{2} + \frac{1}{3} + \cdots + \frac{1}{2n+1} \right]$$

and these sums become infinite as $n \to \infty$ so that $f(x)$ cannot be of bounded variation in $[0, 2/\pi]$. However $f(x)$ is continuous [see Problem 1.40, page 19].

6.6. Prove Theorem 6-7, page 96: If $f(x)$ is of bounded variation in $[a, b]$, then it is bounded in $[a, b]$.

Choose as subdivision points of $[a, b]$ the points a, x, b where $a \leq x \leq b$. Then since $f(x)$ is of bounded variation, there exists a constant M such that

$$|f(x) - f(a)| + |f(b) - f(x)| \leq M$$

and so
$$|f(x) - f(a)| \leq M$$

Then $\quad |f(x)| = |f(x) - f(a) + f(a)| \leq |f(x) - f(a)| + |f(a)| \leq M + |f(a)|$

i.e. $f(x)$ is bounded.

6.7. Prove that the sum of two functions of bounded variation is also of bounded variation.

Let $f(x)$ and $g(x)$ be the functions and $h(x) = f(x) + g(x)$. Then since $f(x)$ and $g(x)$ are of bounded variation, there exist constants M_1 and M_2 such that for all possible partitions

$$\sum_{k=1}^{n} |f(x_k) - f(x_{k-1})| \leq M_1, \qquad \sum_{k=1}^{n} |g(x_k) - g(x_{k-1})| < M_2$$

Then since $\quad |h(x_k) - h(x_{k-1})| \leq |f(x_k) - f(x_{k-1})| + |g(x_k) - g(x_{k-1})|$

we have

$$\sum_{k=1}^{n} |h(x_k) - h(x_{k-1})| \leq \sum_{k=1}^{n} |f(x_k) - f(x_{k-1})| + \sum_{k=1}^{n} |g(x_k) - g(x_{k-1})| \leq M_1 + M_2$$

so that $h(x) = f(x) + g(x)$ is of bounded variation.

6.8. Prove that the product of two functions of bounded variation is also of bounded variation.

We have

$$|f(x_k)\,g(x_k) - f(x_{k-1})\,g(x_{k-1})| = |f(x_k)\,g(x_k) - f(x_{k-1})\,g(x_k) + f(x_{k-1})\,g(x_k) - f(x_{k-1})\,g(x_{k-1})|$$

$$\leq |g(x_k)|\,|f(x_k) - f(x_{k-1})| + |f(x_{k-1})|\,|g(x_k) - g(x_{k-1})|$$

$$\leq P\,|f(x_k) - f(x_{k-1})| + Q\,|g(x_k) - g(x_{k-1})|$$

where P and Q are upper bounds of $g(x)$ and $f(x)$ which exist by Problem 6.6. Then summing from $k = 1$ to n we have

$$\sum_{k=1}^{n} |f(x_k)\,g(x_k) - f(x_{k-1})\,g(x_{k-1})| \leq P \sum_{k=1}^{n} |f(x_k) - f(x_{k-1})| + Q \sum_{k=1}^{n} |g(x_k) - g(x_{k-1})|$$

$$\leq PM_1 + QM_2$$

since $f(x)$ and $g(x)$ are of bounded variation. Thus $f(x)\,g(x)$ is of bounded variation.

6.9. Let $f(x) = G(x) - H(x)$ where $G(x)$ and $H(x)$ are bounded and monotonic increasing in $[a, b]$. Prove that $f(x)$ is of bounded variation in $[a, b]$.

We have

$$|f(x_k) - f(x_{k-1})| = |[G(x_k) - H(x_k)] - [G(x_{k-1}) - H(x_{k-1})]|$$

$$= |[G(x_k) - G(x_{k-1})] - [H(x_k) - H(x_{k-1})]|$$

$$\leq |G(x_k) - G(x_{k-1})| + |H(x_k) - H(x_{k-1})|$$

$$= G(x_k) - G(x_{k-1}) + H(x_k) - H(x_{k-1})$$

since $G(x_k) \geq G(x_{k-1})$, $H(x_k) \geq H(x_{k-1})$. Then

$$\sum_{k=1}^{n} |f(x_k) - f(x_{k-1})| \leq \sum_{k=1}^{n} [G(x_k) - G(x_{k-1})] + \sum_{k=1}^{n} [H(x_k) - H(x_{k-1})]$$

$$= G(b) - G(a) + H(b) - H(a)$$

from which the required result follows.

We can also prove the result if $G(x)$ and $H(x)$ are bounded and monotonic decreasing.

6.10. Prove that if $f(x)$ is of bounded variation in $[a, b]$, then $f(x)$ can be written as the difference of two monotonic increasing functions $G(x)$ and $H(x)$, i.e. $f(x) = G(x) - H(x)$.

Let

$$V = \sum_{k=1}^{n} |f(x_k) - f(x_{k-1})| \tag{1}$$

and let us denote the differences $f(x_k) - f(x_{k-1})$ briefly by Δf. We can decompose the sum (1) into terms where $\Delta f > 0$ and those where $\Delta f < 0$. Then we can write (1) as

$$V = \sum_{\Delta f > 0} |\Delta f| + \sum_{\Delta f < 0} |\Delta f|$$

$$= \sum_{\Delta f > 0} \Delta f - \sum_{\Delta f < 0} \Delta f$$

and so we have

$$V = P + N \tag{2}$$

where

$$P = \sum_{\Delta f > 0} \Delta f, \qquad N = -\sum_{\Delta f < 0} \Delta f \tag{3}$$

Now

$$P - N = \sum_{\Delta f > 0} \Delta f + \sum_{\Delta f < 0} \Delta f = \sum_{k=1}^{n} [f(x_k) - f(x_{k-1})] = f(b) - f(a) \tag{4}$$

From (2) and (4) we find

$$V = 2P + f(a) - f(b), \qquad V = 2N + f(b) - f(a) \tag{5}$$

Let $\mathcal{V}_a^b, \mathcal{P}_a^b, \mathcal{N}_a^b$ denote the least upper bounds of V, P, N in $[a, b]$ respectively. Then from (5) we have

$$\mathcal{V}_a^b = 2\mathcal{P}_a^b + f(a) - f(b), \qquad \mathcal{V}_a^b = 2\mathcal{N}_a^b + f(b) - f(a) \tag{6}$$

These results hold for the interval $[a, b]$. Now if we consider the interval $[a, x]$ we have similarly

$$\mathcal{V}_a^x = 2\mathcal{P}_a^x + f(a) - f(x), \qquad \mathcal{V}_a^x = 2\mathcal{N}_a^x + f(x) - f(a) \tag{7}$$

Then solving for $f(x)$ in (7) we find

$$f(x) = f(a) + \mathcal{P}_a^x - \mathcal{N}_a^x$$

Since $\mathcal{P}_a^x$ and $\mathcal{N}_a^x$ are monotonic increasing functions of x we can write

$$G(x) = f(a) + \mathcal{P}_a^x, \qquad H(x) = \mathcal{N}_a^x$$

so that as required

$$f(x) = G(x) - H(x)$$

where $G(x)$ and $H(x)$ are monotonic increasing.

We call $\mathcal{V}_a^x$ the *total variation* of $f(x)$ in $[a, x]$, $\mathcal{P}_a^x$ the *positive variation* of $f(x)$ in $[a, x]$ and $\mathcal{N}_a^x$ the negative variation of $f(x)$ in $[a, x]$.

DERIVATIVES OF MONOTONIC FUNCTIONS

6.11. Let $f(x)$ be a monotonic increasing function in $[a, b]$. Prove that

$$E = \{x : D^+ f(x) > D_+ f(x)\}$$

has measure zero.

Consider the sets

$$E_{rs} = \{x : D_+ f(x) < r < s < D^+ f(x)\}$$

where r and s are any rational numbers. It is clear that $E = \cup E_{rs}$ where the union is taken over all pairs of rational numbers r, s. Since the rational numbers are countable, we have by Problem 2.4

$$m_e(E) \leq \sum m_e(E_{rs})$$

where the sum is taken over all pairs of rational numbers r, s.

The required result will be proved if we can show that for each fixed pair of rational numbers r, s we have $m_e(E_{rs}) = 0$.

Let $m_e(E_{rs}) = \kappa$. Then given $\epsilon > 0$ there is an open set $O \supset E_{rs}$ such that $m_e(E_{rs}) > m(O) - \epsilon$, i.e. $m(O) < \kappa + \epsilon$.

Now since $D_+ f(x) < r$, it follows that corresponding to every point of E_{rs} there is an interval $[x, x + h]$ in O whose length is arbitrarily small such that

$$\frac{f(x + h) - f(x)}{h} < r \qquad \text{or} \qquad f(x + h) - f(x) < rh$$

By the Vitali covering theorem we can choose a set of n disjoint intervals $I_k = [x_k, x_k + h_k]$, $k = 1, \ldots, n$, such that if $A = \bigcup_{k=1}^n I_k$,

$$m_e(E_{rs} - A) < \epsilon \tag{1}$$

Summing over these intervals we find

$$\sum_{k=1}^n [f(x_k + h_k) - f(x_k)] < r \sum_{k=1}^n h_k \leq rm(O) \leq r(\kappa + \epsilon) \tag{2}$$

Let us now consider the set B consisting of all points of $A = \bigcup_{k=1}^n I_k$ except for the end points. Since $m_e(E_{rs} - B) = m_e(E_{rs} - A)$, it follows from (1) that

$$m_e(E_{rs} - B) < \epsilon \tag{3}$$

Denote by C the set of all points common to E_{rs} and B, i.e. $C = E_{rs} \cap B$. There will be a set of intervals $[y, y+l]$ contained in B whose lengths are arbitrarily small. In view of the fact that $D^+ f(x) > s$, we have for this set of intervals

$$\frac{f(y+l) - f(y)}{l} > s \quad \text{or} \quad f(y+l) - f(y) > sl \tag{4}$$

Now since this set of intervals is a Vitali covering of C we can choose a set of disjoint intervals $J_p = [y_p, y_p + l_p]$, $p = 1, \ldots, q$, for which (4) holds, i.e.

$$f(y_p + l_p) - f(y_p) > sl_p, \qquad p = 1, \ldots, q \tag{5}$$

and such that if $G = \bigcup_{p=1}^{q} J_p$,

$$m_e(C - G) < \epsilon \tag{6}$$

Summing both sides of (5) from $p = 1$ to q we have

$$\sum_{p=1}^{q} [f(y_p + l_p) - f(y_p)] > s \sum_{p=1}^{q} l_p \tag{7}$$

Now since $f(x)$ is monotonic increasing and each interval J_p is contained in some interval I_k, we have

$$\sum_{p=1}^{q} [f(y_p + l_p) - f(y_p)] \leqq \sum_{k=1}^{n} [f(x_k + h_k) - f(x_k)] \tag{8}$$

Then from (2) and (8),

$$s \sum_{p=1}^{q} l_p < \sum_{p=1}^{q} [f(y_p + l_p) - f(y_p)] \leqq \sum_{k=1}^{n} [f(x_k + h_k) - f(x_k)] \leqq r(\kappa + \epsilon) \tag{9}$$

Now since $\sum_{p=1}^{q} l_p = m(G)$, we see from (9) that

$$m(G) \leqq \frac{r(\kappa + \epsilon)}{s}$$

Then we have
$$\begin{aligned} \kappa = m_e(E_{rs}) &\leqq m_e(E_{rs} \cap B) + m_e(E_{rs} \cap \widetilde{B}) \\ &= m_e(C) + m_e(E_{rs} - B) \\ &\leqq m_e(C) + \epsilon \\ &\leqq m_e(C \cap G) + m_e(C \cap \widetilde{G}) + \epsilon \\ &= m_e(G) + m_e(C - G) + \epsilon \\ &\leqq \frac{r(\kappa + \epsilon)}{s} + 2\epsilon \end{aligned}$$

Solving the inequality for κ, we find

$$0 \leqq \kappa < \frac{(r + 2s)\epsilon}{s - r}$$

Since ϵ can be made arbitrarily small, it follows that $\kappa = m_e(E_{rs}) = 0$ as required.

6.12. Let $f(x)$ be a monotonic increasing function in $[a, b]$. Prove that $S = \{x : D^+ f(x) = \infty\}$ has measure zero.

Suppose the contrary, i.e. $m_e(S) = \lambda > 0$. Then if K is any positive number, we can find for each $x \in S$ a set of intervals $[x, x+h]$ such that

$$\frac{f(x+h) - f(x)}{h} > K$$

Now by Vitali's covering theorem there exists a disjoint set of intervals $I_k = [x_k, x_k + h_k]$, $k = 1, \ldots, n$ such that if $A = \bigcup_{k=1}^{n} I_k$,

$$m_e(S - A) < \epsilon \tag{1}$$

For this set of intervals we have

$$\frac{f(x_k + h_k) - f(x_k)}{h_k} > K$$

or

$$h_k < \frac{f(x_k + h_k) - f(x_k)}{K}$$

Then summing from $k = 1$ to n we have

$$m_e(A) = \sum_{k=1}^{n} h_k < \frac{1}{K} \sum_{k=1}^{n} [f(x_k + h_k) - f(x_k)] \leq \frac{f(b) - f(a)}{K}$$

where we have used the fact that $f(x)$ is monotonic increasing in $[a, b]$.

Thus we can choose K so large that

$$m_e(A) < \epsilon \tag{2}$$

and from (1) and (2) we have

$$m_e(S) \leq m_e(S - A) + m_e(A) < 2\epsilon$$

Since ϵ can be made arbitrarily small, we have $m_e(S) = 0$ or $m(S) = 0$.

6.13. Prove that a monotonic increasing function has a derivative almost everywhere.

In Problem 6.11 it was shown that the measure of the set $\{x : D_+ f(x) < D^+ f(x)\}$ is zero. In the same manner we can show [see Problems 6.55, 6.56, 6.57] that the measures of other sets such as

$$\{x : D_+ f(x) = D^+ f(x)\}, \quad \{x : D_- f(x) < D^- f(x)\}, \quad \{x : D_- f(x) < D^+ f(x)\}, \quad \{x : D_+ f(x) < D^- f(x)\}$$

and so on are also zero.

It follows from these results that

$$D_+ f = D_- f = D^+ f = D^- f$$

almost everywhere, so that the derivative exists [i.e. is finite] almost everywhere.

6.14. Prove that a monotonic decreasing function has a derivative almost everywhere.

If $f(x)$ is monotonic increasing, then $-f(x)$ is monotonic decreasing. Then since $D f(x)$ exists almost everywhere, so also does $D[-f(x)] = -D f(x)$ exist almost everywhere.

6.15. Prove that a function of bounded variation has a derivative almost everywhere.

This follows at once from the fact that a function of bounded variation can be written as the difference between two monotonic increasing functions which by Problem 6.13 have derivatives almost everywhere.

ABSOLUTE CONTINUITY

6.16. Prove Theorem 6-15, page 98: An absolutely continuous function is continuous.

If $f(x)$ is absolutely continuous, then given $\epsilon > 0$ we can find $\delta > 0$ such that

$$\sum_{k=1}^{n} |f(x_k + h_k) - f(x_k)| < \epsilon \quad \text{whenever} \quad \sum_{k=1}^{n} |h_k| < \delta$$

By taking $n = 1$ as a special case we have

$$|f(x + h) - f(x)| < \epsilon \quad \text{whenever} \quad |h| < \delta$$

so that $f(x)$ is continuous.

6.17. If $f(x)$ and $g(x)$ are absolutely continuous, prove that $f(x) + g(x)$ is absolutely continuous.

Since $f(x)$ and $g(x)$ are absolutely continuous, given $\epsilon > 0$ there exists $\delta > 0$ such that

$$\sum_{k=1}^{n} |f(x_k + h_k) - f(x_k)| < \frac{\epsilon}{2} \quad \text{whenever} \quad \sum_{k=1}^{n} |h_k| < \delta$$

$$\sum_{k=1}^{n} |g(x_k + h_k) - g(x_k)| < \frac{\epsilon}{2} \quad \text{whenever} \quad \sum_{k=1}^{n} |h_k| < \delta$$

Then since

$$|[f(x_k + h_k) + g(x_k + h_k)] - [f(x_k) + g(x_k)]| \leqq |f(x_k + h_k) - f(x_k)| + |g(x_k + h_k) - g(x_k)|$$

we have for $\sum |h_k| < \delta$,

$$\sum_{k=1}^{n} |[f(x_k + h_k) + g(x_k + h_k)] - [f(x_k) + g(x_k)]|$$
$$\leqq \sum_{k=1}^{n} |f(x_k + h_k) - f(x_k)| + \sum_{k=1}^{n} |g(x_k + h_k) - g(x_k)|$$
$$< \frac{\epsilon}{2} + \frac{\epsilon}{2} = \epsilon$$

which proves the required result.

6.18. Prove Theorem 6-17, page 98: An absolutely continuous function is of bounded variation.

If $f(x)$ is absolutely continuous, then given $\epsilon > 0$ there exists $\delta > 0$ such that

$$\sum_{k=1}^{n} |f(x_k + h_k) - f(x_k)| < \epsilon \quad \text{whenever} \quad \sum |h_k| < \delta$$

But this implies that the total variation over any interval of length δ does not exceed ϵ. Thus the total variation over the interval $[a, b]$, which is of length $b - a$, does not exceed $(b - a)\epsilon/\delta$. It follows that the total variation of $f(x)$ in $[a, b]$ is bounded, i.e. $f(x)$ is of bounded variation.

6.19. Prove Theorem 6-18, page 98: An absolutely continuous function has a derivative almost everywhere.

This follows at once from Theorem 6-17 and Theorem 6-13.

6.20. Prove that there are continuous functions which are not absolutely continuous.

If a continuous function were absolutely continuous, then by Theorem 6-18, page 98, it would have a derivative almost everywhere. However, there exist continuous functions which have derivatives nowhere. See for example Problem 6.66.

It is true that even continuous functions which are of bounded variation need not be absolutely continuous.

6.21. Prove Theorem 6-19, page 98: If $f(x)$ is absolutely continuous in $[a, b]$ and $f'(x) = 0$ almost everywhere in $[a, b]$, then $f(x)$ is a constant in $[a, b]$.

Denote by E the set where $f'(x) = 0$. For each $x \in E$ there exists an arbitrarily small interval $[x, x + h]$ for which

$$\left| \frac{f(x + h) - f(x)}{h} \right| < \epsilon \quad \text{i.e.} \quad |f(x + h) - f(x)| < \epsilon h \qquad (1)$$

Then by Vitali's covering theorem [see page 33] there exists a finite set of disjoint intervals $I_k = [x_k, x_k + h_k]$, $k = 1, \ldots n$, which cover E, and thus $[a, b]$ except for a set of measure less than δ. We have

$$x_0 = a \leqq x_1 < x_1 + h_1 \leqq x_2 < x_2 + h_2 \leqq \cdots \leqq x_n < x_n + h_n \leqq x_{n+1} = b$$

Then
$$f(b) - f(a) = f(x_{n+1}) - f(x_n + h_n) + f(x_n + h_n) - f(x_n)$$
$$+ \cdots + f(x_1 + h_1) - f(x_1) + f(x_1) - f(x_0)$$
$$= \sum_{k=0}^{n} [f(x_{k+1}) - f(x_k + h_k)] + \sum_{k=1}^{n} [f(x_k + h_k) - f(x_k)]$$

so that
$$|f(b) - f(a)| \leqq \sum_{k=0}^{n} |f(x_{k+1}) - f(x_k + h_k)| + \sum_{k=1}^{n} |f(x_k + h_k) - f(x_k)| \qquad (2)$$

Now by the absolute continuity of $f(x)$ we have

$$\sum_{k=0}^{n} |f(x_{k+1}) - f(x_k + h_k)| < \epsilon_1 \quad \text{for} \quad |x_{k+1} - (x_k + h_k)| < \delta$$

Also from (1) we have

$$\sum_{k=1}^{n} |f(x_k + h_k) - f(x_k)| \leqq \epsilon \sum_{k=1}^{n} h_k \leqq \epsilon(b - a)$$

Thus (2) becomes
$$|f(b) - f(a)| \leqq \epsilon(b - a) + \epsilon_1$$

and since ϵ and ϵ_1 can be made arbitrarily small we have $f(b) = f(a)$. Similarly we can show that for any x, $f(x) = f(a)$, i.e. $f(x)$ is constant in $[a, b]$.

THEOREMS ON INDEFINITE LEBESGUE INTEGRALS

6.22. Let $f(x)$ be Lebesgue integrable in $[a, b]$. If $F(x) = \int_a^x f(u)\, du$, prove that $F(x)$ is continuous in $[a, b]$.

We have
$$F(x + h) - F(x) = \int_a^{x+h} f(u)\, du - \int_a^x f(u)\, du$$

or
$$F(x + h) - F(x) = \int_x^{x+h} f(u)\, du$$

Then using Theorem 5-15, page 74,
$$\lim_{h \to 0} [F(x + h) - F(x)] = \lim_{h \to 0} \int_x^{x+h} f(u)\, du = 0$$

so that
$$\lim_{h \to 0} F(x + h) = F(x)$$

and $F(x)$ is continuous.

6.23. If $f(x)$ is Lebesgue integrable in $[a, b]$, prove that
$$F(x) = \int_a^x f(u)\, du$$
is of bounded variation in $[a, b]$.

Subdivide $[a, b]$ using the points of subdivision $a = x_0 < x_1 < \cdots < x_n = b$. Then

$$|F(x_k) - F(x_{k-1})| = \left| \int_{x_{k-1}}^{x_k} f(u)\, du \right| \leqq \int_{x_{k-1}}^{x_k} |f(u)|\, du$$

Then summing from $k = 1$ to n, we have

$$\sum_{k=1}^{n} |F(x_k) - F(x_{k-1})| \leq \sum_{k=1}^{n} \int_{x_{k-1}}^{x_k} |f(u)|\, du = \int_a^b |f(u)|\, du$$

But since $f(x)$ is integrable, it is absolutely integrable, i.e. the last integral is finite and so $F(x)$ is of bounded variation.

For an alternative proof see Problem 6.67.

6.24. Prove Theorem 6-22, page 98: If $g(x)$ is integrable on $[a, b]$ and

$$\int_a^x g(u)\, du = 0$$

for all x in $[a, b]$, then $g(x) = 0$ almost everywhere in $[a, b]$.

Suppose that $g(x) \neq 0$ almost everywhere in $[a, b]$. Then we must have either $g(x) > 0$ or $g(x) < 0$ on some set of measure greater than zero.

Consider first the case where $g(x) > 0$ on some set of measure greater than zero. From the fact that

$$\int_a^x g(u)\, du = 0$$

we see that the integral of $g(x)$ on any open interval is zero. Now any open set E can be expressed as a countable union of disjoint open intervals $E_1, E_2, \ldots$. Since

$$\int_{E_k} g(x)\, dx = 0$$

we have

$$\int_E g(x)\, dx = \sum_{k=1}^{\infty} \int_{E_k} g(x)\, dx = 0$$

Thus if we assume $g(x) > 0$ on E, we arrive at a contradiction in view of Theorem 5-13, page 74.

Similarly if $g(x) < 0$ we arrive at a contradiction. Thus $g(x) = 0$ almost everywhere.

6.25. Prove Theorem 6-23, page 98: If $g(x)$ is continuous and monotonic increasing in $[a, b]$, then $g'(x)$ is integrable and

$$\int_a^b g'(x)\, dx \leq g(b) - g(a)$$

Since $g(x)$ is monotonic increasing we have by Problem 6.13,

$$\lim_{h \to 0} \frac{g(x+h) - g(x)}{h} = g'(x) \qquad \text{exists almost everywhere}$$

Then by Fatou's theorem it follows that

$$\lim_{h \to 0} \int_a^b \frac{g(x+h) - g(x)}{h}\, dx \geq \int_a^b g'(x)\, dx$$

But the left side is equal to

$$\lim_{h \to 0} \left\{ \frac{1}{h} \int_a^b g(x+h)\, dx - \frac{1}{h} \int_a^b g(x)\, dx \right\} = \lim_{h \to 0} \left\{ \frac{1}{h} \int_{a+h}^{b+h} g(x)\, dx - \frac{1}{h} \int_a^b g(x)\, dx \right\}$$

$$= \lim_{h \to 0} \left\{ \frac{1}{h} \int_{a+h}^a g(x)\, dx + \frac{1}{h} \int_b^{b+h} g(x)\, dx \right\}$$

$$= \lim_{h \to 0} \left\{ \frac{1}{h} \int_b^{b+h} g(x)\, dx - \frac{1}{h} \int_a^{a+h} g(x)\, dx \right\}$$

$$= g(b) - g(a)$$

and so the required result follows.

6.26. Prove Theorem 6-24, page 98, for the case where $f(x)$ is bounded.

If $f(x)$ is bounded, i.e. $|f(x)| \leqq M$, then

$$\left| \frac{F(x+h) - F(x)}{h} \right| \;=\; \left| \frac{1}{h} \int_x^{x+h} f(u)\, du \right| \;\leqq\; M$$

Letting $h \to 0$ we find, since the left side approaches $F'(x)$ almost everywhere, that

$$|F'(x)| \;\leqq\; M$$

Then by the theorem of bounded convergence we have

$$\lim_{h \to 0} \int_a^x \frac{F(u+h) - F(u)}{h}\, du \;=\; \int_a^x F'(u)\, du \qquad (1)$$

Now

$$\int_a^x \frac{F(u+h) - F(u)}{h}\, du \;=\; \frac{1}{h} \int_a^x F(u+h)\, du \;-\; \frac{1}{h} \int_a^x F(u)\, du$$

$$=\; \frac{1}{h} \int_{a+h}^{x+h} F(u)\, du \;-\; \frac{1}{h} \int_a^x F(u)\, du$$

$$=\; \frac{1}{h} \int_{a+h}^{a} F(u)\, du \;+\; \frac{1}{h} \int_a^x F(u)\, du \;+\; \frac{1}{h} \int_x^{x+h} F(u)\, du \;-\; \frac{1}{h} \int_a^x F(u)\, du$$

$$=\; \frac{1}{h} \int_x^{x+h} F(u)\, du \;-\; \frac{1}{h} \int_a^{a+h} F(u)\, du$$

Thus

$$\lim_{h \to 0} \int_a^x \frac{F(u+h) - F(u)}{h}\, du \;=\; \lim_{h \to 0} \frac{1}{h} \int_x^{x+h} F(u)\, du \;-\; \lim_{h \to 0} \frac{1}{h} \int_a^{a+h} F(u)\, du$$

$$=\; F(x) - F(a)$$

Then using (1) we have

$$\int_a^x F'(u)\, du \;=\; F(x) - F(a) \;=\; \int_a^x f(u)\, du$$

so that for all x,

$$\int_a^x [F'(u) - f(u)]\, du \;=\; 0$$

Then by Theorem 5-13, page 74, we have $F'(x) = f(x)$ almost everywhere.

6.27. Prove Theorem 6-24, page 98, for the case where $f(x)$ is integrable but unbounded.

It is sufficient to prove the result for $f(x) \geqq 0$ since it will then be true for $f^+(x)$ and $f^-(x)$ and can be extended to the case where $f(x)$ has arbitrary sign as in Chapter 5 by using the fact that $f(x) = f^+(x) - f^-(x)$.

We have

$$[f(x)]_p \;=\; \begin{cases} f(x), & \text{for } f(x) \leqq p \\ p, & \text{for } f(x) > p \end{cases}$$

Then since $[f(x)]_p \leqq f(x)$ so that $f(x) - [f(x)]_p \geqq 0$,

we have

$$\int_a^x \{ f(u) - [f(u)]_p \}\, du \;\geqq\; 0$$

Thus

$$\frac{d}{dx} \int_a^x \{ f(u) - [f(u)]_p \}\, du \;\geqq\; 0$$

or

$$\frac{d}{dx} \int_a^x f(u)\, du \;\geqq\; \frac{d}{dx} \int_a^x [f(u)]_p\, du$$

i.e.

$$\frac{d}{dx} F(x) \;=\; F'(x) \;\geqq\; [f(x)]_p \qquad \text{almost everywhere}$$

Taking the limit as $p \to \infty$, we see that $F'(x) \geqq f(x)$ almost everywhere so that on integrating we have

$$\int_a^b F'(x)\,dx \;\; \geqq \;\; \int_a^b f(x)\,dx \;\; = \;\; F(b) - F(a) \tag{1}$$

However by Problem 6.25,

$$\int_a^b F'(x)\,dx \;\; \leqq \;\; F(b) - F(a) \tag{2}$$

It follows from (1) and (2) that

$$\int_a^b F'(x)\,dx \;\; = \;\; F(b) - F(a) \;\; = \;\; \int_a^b f(x)\,dx$$

i.e. $$\int_a^b [F'(x) - f(x)]\,dx \;\; = \;\; 0$$

or replacing b by x we have

$$\int_a^x [F'(u) - f(u)]\,du \;\; = \;\; 0$$

for all x in $[a, b]$. Then since the integrand is non-negative it follows from Theorem 5-13, page 74, that $F'(x) = f(x)$ almost everywhere.

6.28. Prove that if $F(x)$ is an integral, then it is absolutely continuous.

We have $$F(x) \;\; = \;\; \int_a^x f(u)\,du$$

Then $$F(x_k + h_k) - F(x_k) \;\; = \;\; \int_a^{x_k + h_k} f(u)\,du - \int_a^{x_k} f(u)\,du \;\; = \;\; \int_{x_k}^{x_k + h_k} f(u)\,du$$

Thus $$|F(x_k + h_k) - F(x_k)| \;\; \leqq \;\; \int_{x_k}^{x_k + h_k} |f(u)|\,du$$

Summing from $k = 1$ to n,

$$\sum_{k=1}^n |F(x_k + h_k) - F(x_k)| \;\; \leqq \;\; \sum_{k=1}^n \int_{x_k}^{x_k + h_k} |f(u)|\,du \;\; = \;\; \int_E |f(u)|\,du$$

where E is the union of the intervals $E_k = [x_k, x_k + h_k]$, $k = 1, 2, \ldots, n$. Then using Theorem 5-15 we have for any given $\epsilon > 0$,

$$\sum_{k=1}^n |F(x_k + h_k) - F(x_k)| \;\; < \;\; \epsilon \quad \text{whenever} \quad m(E) \;\; = \;\; \sum_{k=1}^n |h_k| \;\; < \;\; \delta$$

i.e. $F(x)$ is absolutely continuous.

6.29. Prove that a function $F(x)$ is an integral if it is absolutely continuous.

Since $F(x)$ is absolutely continuous, it is continuous and of bounded variation by Problems 6.16 and 6.18. Thus there are continuous monotonic increasing functions $G(x)$ and $H(x)$ such that

$$F(x) \;\; = \;\; G(x) - H(x)$$

[see Problem 6.51] and we have almost everywhere

$$F'(x) \;\; = \;\; G'(x) - H'(x)$$

Now from Problem 6.25, $G'(x)$ and $H'(x)$ are integrable and thus $F'(x)$ is integrable, i.e.

$$\int_a^x F'(u)\,du \tag{1}$$

exists. But the integral (1) is absolutely continuous [Problem 6.28] and thus

$$g(x) \;=\; \int_a^x F'(u)\,du \;-\; F(x) \tag{2}$$

is absolutely continuous.   Then we have almost everywhere

$$g'(x) \;=\; \frac{d}{dx}\int_a^x F'(u)\,du \;-\; F(x) \;=\; F'(x) \;-\; F'(x) \;=\; 0$$

Thus from Problem 6.21, $g(x) = c$, a constant, and from (2) it follows that $c = -F(a)$ so that (2) can be written

$$F(x) \;=\; \int_a^x F'(u)\,du \;+\; F(a)$$

i.e. $F(x)$ is an integral.

6.30. Prove Theorem 6-25, page 99:   If $F(x)$ is absolutely continuous, then

$$\int_a^b F'(x)\,dx \;=\; F(b) \;-\; F(a)$$

This follows at once from Equation (2) of Problem 6.29, i.e.

$$\int_a^x F'(u)\,du \;=\; F(x) \;-\; F(a)$$

so that if $x = b$, $\qquad \displaystyle\int_a^b F'(x)\,dx \;=\; \int_a^b F'(u)\,du \;=\; F(b) \;-\; F(a)$

INTEGRATION BY PARTS

6.31. Prove Theorem 6-27, page 99:   Let $F(x)$ and $G(x)$ be indefinite integrals of the integrable functions $f(x)$ and $g(x)$ respectively.   Then

$$\int_a^b F(x)\,g(x)\,dx \;=\; F(x)\,G(x)\Big|_a^b \;-\; \int_a^b f(x)\,G(x)\,dx$$

Since an integral is absolutely continuous [Problem 6.28], it follows that $F(x)\,G(x)$ is absolutely continuous [Problem 6.61].   Then by Problem 6.19

$$\frac{d}{dx}\,[F(x)\,G(x)] \;=\; F'(x)\,G(x) \;+\; F(x)\,G'(x) \;=\; f(x)\,G(x) \;+\; F(x)\,g(x)$$

exists almost everywhere and, by Problem 6.30,

$$\int_a^b \frac{d}{dx}\,[F(x)\,G(x)]\,dx \;=\; F(x)\,G(x)\Big|_a^b$$

so that $\qquad \displaystyle\int_a^b [f(x)\,G(x) + F(x)\,g(x)]\,dx \;=\; F(x)\,G(x)\Big|_a^b$

Thus $\qquad \displaystyle\int_a^b F(x)\,g(x)\,dx \;=\; F(x)\,G(x)\Big|_a^b \;-\; \int_a^b f(x)\,G(x)\,dx$

Supplementary Problems

MONOTONIC FUNCTIONS AND FUNCTIONS OF BOUNDED VARIATION

6.32. Prove a result analogous to that of Problem 6.1 if $f(x)$ is monotonic decreasing in $[a, b]$.

6.33. Prove Theorem 6-4, page 96, for the case where $f(x)$ is monotonic decreasing in $[a, b]$.

6.34. Prove Theorem 6-5, page 96, for the case where $f(x)$ is monotonic decreasing.

6.35. If $f(x)$ and $g(x)$ are monotonic increasing functions in $[a, b]$, determine which of the following functions are also monotonic increasing in $[a, b]$: (a) $f(x) + g(x)$, (b) $f(x) - g(x)$, (c) $f(x)\,g(x)$, (d) $f(x)/g(x)$, $g(x) \neq 0$.

6.36. Work Problem 6.35 if (a) $f(x)$ and $g(x)$ are monotonic decreasing, (b) $f(x)$ is monotonic increasing but $g(x)$ is monotonic decreasing.

6.37. Prove that the difference of two functions of bounded variation is also of bounded variation.

6.38. Prove that the quotient of two functions of bounded variation is also of bounded variation provided that there is no division by zero.

6.39. Work Problem 6.9, page 102, if $G(x)$ and $H(x)$ are bounded and monotonic decreasing in $[a, b]$.

6.40. Are the results of Problems 6.9 and 6.39 valid if the word bounded is not included?   Explain.

6.41. Is the converse of Theorem 6-7 true?   Explain.

6.42. If $f(x)$ and $g(x)$ are of bounded variation, prove that (a) $f(x) - g(x)$ and (b) $f(x)/g(x)$, $g(x) \neq 0$ are also of bounded variation.   [*Hint*: For (b) first show that $1/g(x)$, $g(x) \neq 0$ is of bounded variation.]

6.43. Prove that $f(x) = x^4 - 3x^3 + 2x^2 - 5x - 6$ is of bounded variation in (a) the interval $[1, 2]$, (b) the interval $[-3, 3]$, (c) any finite interval.

6.44. Prove that a polynomial is of bounded variation in any finite interval.

6.45. Prove that $f(x) = \begin{cases} \cos x, & 0 \leq x < \pi \\ \sin x, & \pi \leq x \leq 2\pi \end{cases}$ is of bounded variation in $[0, 2\pi]$.

6.46. Prove that $f(x) = \begin{cases} x^2 \sin(1/x), & x \neq 0 \\ 0, & x = 0 \end{cases}$ is of bounded variation in $[0, 2/\pi]$.

6.47. Prove Theorem 6-10.

6.48. Let $f(x)$ defined in $[a, b]$ satisfy the condition $|f(x_1) - f(x_2)| \leq K|x_1 - x_2|$ where K is a constant. Prove that $f(x)$ is of bounded variation in $[a, b]$.

6.49. Prove Theorem 6-11.   [*Hint*: Use the law of the mean.]

6.50. If $f(x)$ is of bounded variation and continuous at $x = x_0$, prove that the total variation $\mathcal{V}_a^x$ is also continuous at x_0.

6.51. Prove that a function is continuous and of bounded variation if and only if it can be expressed as the difference of two monotonic increasing (or monotonic decreasing) continuous functions.

6.52. Prove that $\mathcal{V}_a^b = \mathcal{V}_a^c + \mathcal{V}_c^b$ where c is a point in $[a, b]$.

DERIVATES AND DERIVATIVES OF A FUNCTION

6.53. Let $f(x) = \begin{cases} x \sin(1/x), & x \neq 0 \\ 0, & x = 0 \end{cases}$. Prove that at $x = 0$ (a) $D^+ f(x) = 1$, (b) $D^- f(x) = 1$,
(c) $D_+ f(x) = -1$, (d) $D_- f(x) = -1$.

6.54. Show that if $D^+ f(x)$ is continuous at $x = a$, then the derivative of $f(x)$ exists at $x = a$. Is the result still true if one of the other derivates is used? Explain.

6.55. Let $f(x)$ be a monotonic increasing function in $[a, b]$. Prove that the sets (a) $\{x : D_+ f(x) < D^- f(x)\}$,
(b) $\{x : D_- f(x) < D^- f(x)\}$ and (c) $\{x : D_- f(x) < D^+ f(x)\}$ all have measure zero.

6.56. Let $f(x)$ be a monotonic increasing function in $[a, b]$. Prove that the sets (a) $\{x : D_+ f(x) = D^+ f(x)\}$,
(b) $\{x : D_- f(x) = D^- f(x)\}$ and (c) $\{x : D_+ f(x) = D_- f(x)\}$ all have measure zero.

6.57. Prove that $\{x : D_- f(x) = -\infty\}$ has measure zero if $f(x)$ is monotonic increasing in $[a, b]$.

6.58. Are the results of Problems 6.55, 6.56 and 6.57 still true if $f(x)$ is (a) monotonic decreasing (b) strictly increasing? Explain.

6.59. Prove Theorem 6-14. [*Hint*: First prove the theorem for monotonic increasing functions.]

ABSOLUTE CONTINUITY

6.60. Prove that (a) $f(x) = x^2$ is absolutely continuous in $[0, 1]$ and (b) any polynomial is absolutely continuous in any finite interval.

6.61. If $f(x)$ and $g(x)$ are absolutely continuous in $[a, b]$, prove that (a) $f(x) - g(x)$, (b) $f(x)\,g(x)$ and (c) $f(x)/g(x)$, $g(x) \neq 0$ are absolutely continuous in $[a, b]$.

6.62. If $f(x)$ and $g(x)$ are absolutely continuous in $[a, b]$ and $f'(x) = g'(x)$ almost everywhere in $[a, b]$, prove that if $f(x)$ and $g(x)$ are equal at one point of $[a, b]$ then they are identically equal.

6.63. If $f(x)$ is absolutely continuous in $[a, b]$, prove that $|f(x)|^p$, $p > 0$ is absolutely continuous in $[a, b]$.

6.64. If $f(x)$ has a derivative (a) everywhere, (b) almost everywhere in $[a, b]$, is it absolutely continuous? Justify your statements. Compare Theorem 6-18.

6.65. Prove Theorem 6-20.

6.66. Consider the *Weierstrass function* defined by

$$f(x) = \sum_{k=0}^{\infty} a^k \cos(b^k \pi x) \quad \text{where} \quad 0 < a < 1 \text{ and } b > 0 \text{ is odd}$$

(a) Prove that $f(x)$ is continuous everywhere. (b) Write the series for $[f(x+h) - f(x)]/h$ as $A_n + B_n$ where A_n represents the sum of the first n terms of the series and B_n is the remainder after n terms and prove that

$$\left| \frac{f(x+h) - f(x)}{h} \right| \geq |B_n| - |A_n|$$

where

$$|A_n| < \frac{\pi a^n b^n}{ab - 1}, \quad |B_n| > \tfrac{2}{3} a^n b^n$$

(c) Use the result in (b) to show that if $ab > 1 + 3\pi/2$ then $f'(x)$ does not exist, thus showing that $f(x)$ is continuous everywhere but has a derivative nowhere.

[*Hint*: To prove the inequality for A_n in (b), use the mean-value theorem on $\cos[b^k \pi(x+h)] - \cos[b^k \pi x]$. To prove the inequality for B_n, first let $b^n x = p_n + q_n$ where p_n is an integer and $-1/2 \leq q_n < 1/2$ and prove that if $h = (1 - q_n)/b^n$ then $2b^n/3 \leq 1/h \leq 2b^n$. Then prove that for $k \geq n$,

$$\cos[b^k \pi(x+h)] = (-1)^{p_n + 1}, \quad \cos[b^k \pi x] = (-1)^{p_n + 1} \cos[b^{k-n} \pi q_n]$$

and estimate B_n using the first term of its corresponding series.]

THEOREMS ON INDEFINITE INTEGRALS

6.67. Work Problem 6.23 by showing directly that

$$F(x) = \int_a^x f(u)\, du$$

is the difference between two monotonic increasing functions.

6.68. If $f(x)$ and $g(x)$ are continuous in $[a, b]$ and

$$\int_a^x f(u)\, du = \int_a^x g(u)\, du$$

prove that $f(x) = g(x)$ at all points of $[a, b]$.

6.69. Prove (a) Theorem 6-26, page 99, (b) Theorem 6-28, page 99.

6.70. Prove that $f'(x)$ is integrable in $[a, b]$ if $f(x)$ is of bounded variation.

6.71. If $f'(x)$ is integrable in $[a, b]$ and $f(x)$ is of bounded variation, is it true that

$$\int_a^b f'(x)\, dx = f(b) - f(a)$$

Explain.

6.72. Is there a result corresponding to that of Theorem 6-23 for monotonic decreasing functions? Explain.

6.73. Let $f(x) = \begin{cases} x^2 \sin(1/x^2), & 0 < x \le \sqrt{2/\pi} \\ 0, & x = 0 \end{cases}$. Prove that (a) $f'(x)$ exists everywhere in $[0, \sqrt{2/\pi}\,]$ but (b) $f'(x)$ is not integrable in $[0, \sqrt{2/\pi}\,]$. Discuss the relationship with Theorem 6-26, page 99. [*Hint*: For part (b) use the fact that $f'(x)$ is Lebesgue integrable if and only if $|f'(x)|$ is Lebesgue integrable.]

6.74. If $F(x) = \int_a^x f(u)\, du$ where $f(x)$ is integrable on $[a, b]$, prove that the total variation of $F(x)$ in $[a, b]$ is $\int_a^b |f(u)|\, du$.

6.75. Prove Theorem 6-29 on change of variables, page 99. [*Hint*: Use Problem 6.61(b).]

6.76. State sufficient conditions under which we can write

$$\int_a^b f(x)\, dg(x) = f(x)\, g(x)\Big|_a^b - \int_a^b g(x)\, f'(x)\, dx$$

6.77. Let $f'(x)$ exist everywhere in $[a, b]$ and suppose that $f'(x) = g(x)$ almost everywhere in $[a, b]$. Prove that $f'(x) = g(x)$ everywhere in $[a, b]$ if $g(x)$ is continuous almost everywhere in $[a, b]$.

6.78. Let $f(x)$ be integrable in $[a, b]$ and $g(x) \ge 0$ be monotonic increasing in $[a, b]$. Prove that there is a number η in $[a, b]$ such that

$$\int_a^b f(x)\, g(x)\, dx = g(b-0) \int_\eta^b f(x)\, dx$$

This is often called the *second mean-value theorem*. A generalization of this is given in Problem 6.80.

6.79. If in Problem 6.78 $g(x) > 0$ is monotonic decreasing, prove that

$$\int_a^b f(x)\, g(x)\, dx = g(a+0) \int_a^\eta f(x)\, dx$$

6.80. If in Problem 6.79 $g(x) > 0$ is any monotonic function, prove that

$$\int_a^b f(x)\, g(x)\, dx = g(a+0) \int_a^\eta f(x)\, dx + g(b-0) \int_\eta^b f(x)\, dx$$

Chapter 7

Mean Convergence

L^p SPACES

The space of all functions $f(x)$ for which $|f(x)|^p$, $p \geqq 1$, is Lebesgue integrable on $[a, b]$, i.e.

$$\int_a^b |f(x)|^p \, dx \; < \; \infty$$

is denoted by $L^p[a, b]$, or briefly L^p if the particular interval is not required. In such case we say that $f(x)$ belongs to L^p or briefly $f(x) \in L^p$. If $p = 1$ we denote L^p by L.

Although we shall use the interval $[a, b]$ in this chapter, all results obtained can be restated using any measurable set E instead of $[a, b]$.

HILBERT SPACE

The space L^2 consisting of all functions $f(x)$ for which

$$\int_a^b |f(x)|^2 \, dx \; < \; \infty$$

is called *Hilbert space*. Functions belonging to Hilbert space are often said to be *square integrable*.

IMPORTANT INEQUALITIES

1. **Schwarz's inequality**

$$\left| \int_a^b f(x)\, g(x) \, dx \right| \; \leqq \; \left\{ \int_a^b |f(x)|^2 \, dx \right\}^{1/2} \left\{ \int_a^b |g(x)|^2 \, dx \right\}^{1/2}$$

where $f(x) \in L^2$, $g(x) \in L^2$.

Equality holds if and only if $f(x)/g(x)$ is constant almost everywhere.

2. **Holder's inequality**

$$\left| \int_a^b f(x)\, g(x) \, dx \right| \; \leqq \; \left\{ \int_a^b |f(x)|^p \, dx \right\}^{1/p} \left\{ \int_a^b |g(x)|^q \, dx \right\}^{1/q}$$

where
$$\frac{1}{p} + \frac{1}{q} \; = \; 1, \qquad p > 1$$

and $f(x) \in L^p$, $g(x) \in L^q$.

Equality occurs if and only if $|f(x)|^p / |g(x)|^q$ is constant almost everywhere.

Note that if $p = 2$, $q = 2$ this inequality reduces to Schwarz's inequality.

3. Minkowski's inequality

$$\int_a^b |f(x) + g(x)|^p\, dx \;\leqq\; \left\{\int_a^b |f(x)|^p\, dx\right\}^{1/p} + \left\{\int_a^b |g(x)|^p\, dx\right\}^{1/p}$$

where $p \geqq 1$.

Equality occurs if $f(x)/g(x)$ is constant almost everywhere.

Analogous inequalities exist where sums of real numbers are used in place of integrals [see Problems 7.26, 7.32, 7.45].

THE L^p SPACES AS METRIC SPACES

If we define a *distance* between two functions $f(x)$ and $g(x)$ in L^p, also called a *norm* in L^p, as

$$D(f,g) \;=\; \|f - g\| \;=\; \left\{\int_a^b |f(x) - g(x)|^p\, dx\right\}^{1/p}$$

then we can show that L^p is a metric space [see Problem 7.12]. For example the triangle inequality involving elements $f(x)$, $g(x)$ and $h(x)$ in L^p can be written

$$\|f - g\| \;\leqq\; \|f - h\| + \|h - g\|$$

and is a consequence of Minkowski's inequality.

The functions are often called *points* of the space. Since two functions which are equal almost everywhere have the distance between them equal to zero, we shall assume that they represent the same point in the space.

When it is necessary to specify the space, we shall write $\|f - g\|_p$ rather than $\|f - g\|$.

THEOREMS INVOLVING FUNCTIONS IN L^p SPACES

Theorem 7-1.   If $f(x) \in L^p$, $p \geqq 1$, it also belongs to L, i.e. $L^p \subset L$. More generally if $p > n \geqq 1$, then $L^p \subset L^n$.

Theorem 7-2.   If $g(x) \in L^p$ and $|f(x)| \leqq |g(x)|$, then $f(x) \in L^p$.

Theorem 7-3.   If $f(x) \in L^p$ and $g(x) \in L^p$, then $f(x)\, g(x) \in L^{p/2}$. In particular if $f(x) \in L^2$ and $g(x) \in L^2$, then $f(x)\, g(x) \in L$.

Theorem 7-4.   If $f(x) \in L^p$ and $g(x) \in L^q$ where $\dfrac{1}{p} + \dfrac{1}{q} = 1$, $p > 1$, then $f(x)\, g(x) \in L$.

Theorem 7-5.   If $f(x) \in L^p$ and $g(x) \in L^p$, then $f(x) \pm g(x) \in L^p$.

MEAN CONVERGENCE

Let $\langle f_n(x)\rangle$ be a sequence of functions which belong to $L^p(a, b)$. If there exists a function $f(x) \in L^p$ such that

$$\lim_{n \to \infty} \int_a^b |f_n(x) - f(x)|^p\, dx \;=\; 0 \tag{1}$$

we say that the sequence $\langle f_n(x)\rangle$ *converges in the mean* or is *mean convergent* to $f(x)$ in the space L^p.

In such case we often write (1) as

$$\underset{n \to \infty}{\text{l.i.m.}}\; f_n(x) \;=\; f(x)$$

which is read "the limit in mean of $f_n(x)$ as $n \to \infty$ is $f(x)$".

Equivalently we can say that $f_n(x)$ approaches $f(x)$ in the mean if for every $\epsilon > 0$ there exists a number $n_0 > 0$ such that

$$\|f_n(x) - f(x)\| < \epsilon \quad \text{whenever} \quad n > n_0$$

Theorem 7-6.   If $\underset{n \to \infty}{\text{l.i.m.}} f_n(x)$ exists, then it is unique [see Problem 7.13].

CAUCHY SEQUENCES IN L^p SPACES

The sequence of functions $\langle f_n(x) \rangle$ belonging to L^p is said to be a *Cauchy sequence* if

$$\lim_{\substack{m \to \infty \\ n \to \infty}} \int_a^b |f_m(x) - f_n(x)|^p \, dx = 0 \tag{3}$$

or, in other words, if given $\epsilon > 0$, there exists a number $n_0 > 0$ such that

$$\|f_m - f_n\| = \int_a^b |f_m(x) - f_n(x)|^p \, dx < \epsilon \quad \text{whenever} \quad m > n_0,\ n > n_0$$

Note the analogy with Cauchy sequences of real numbers. We have the following

Theorem 7-7.   If $\langle f_n(x) \rangle$ converges in mean to $f(x)$ in L^p, then $\langle f_n(x) \rangle$ is a Cauchy sequence.

COMPLETENESS OF L^p.   THE RIESZ-FISCHER THEOREM

The space L^p is said to be *complete* if every Cauchy sequence in the space converges in the mean to a function in the space. The following theorem, known as the *Riesz-Fischer theorem*, is of fundamental importance.

Theorem 7-8   [**Riesz-Fischer**].   Any L^p space is complete.   In other words if $\langle f_n(x) \rangle$ is a sequence of functions belonging to L^p and

$$\lim_{\substack{m \to \infty \\ n \to \infty}} \int_a^b |f_m(x) - f_n(x)|^p \, dx = 0$$

i.e. if $\langle f_n(x) \rangle$ is a Cauchy sequence, then there is a function $f(x) \in L^p$ to which $f_n(x)$ converges in the mean.   This function is unique apart from a set of measure zero.

CONVERGENCE IN MEASURE

Let $\langle f_n(x) \rangle$ be a sequence of measurable functions defined almost everywhere.   Then $\langle f_n(x) \rangle$ is said to *converge in measure* to $f(x)$ if

$$\lim_{n \to \infty} m\{x : |f_n(x) - f(x)| \geqq \sigma\} = 0$$

for all $\sigma > 0$.

The following theorems hold.

Theorem 7-9.   If $\langle f_n(x) \rangle$ converges almost everywhere to $f(x)$, then it converges in measure to $f(x)$.   However, the converse is not valid but see Theorem 7-11.

Theorem 7-10.   If $\langle f_n(x) \rangle$ converges in the mean to $f(x)$, then it converges in measure to $f(x)$.

Theorem 7-11.   If $\langle f_n(x) \rangle$ converges in measure to $f(x)$ on a set E, then there exists a subsequence $\langle f_{n_k}(x) \rangle$ which converges almost everywhere to $f(x)$.

Theorem 7-12. If $\langle f_n(x) \rangle$ converges in the mean to $f(x)$ on a set E, then there exists a subsequence $\langle f_{n_k}(x) \rangle$ which converges almost everywhere to $f(x)$.

It is of interest that if $\langle f_n(x) \rangle$ converges to $f(x)$ everywhere, it does not necessarily converge in the mean to $f(x)$. Conversely if $\langle f_n(x) \rangle$ converges in the mean to $f(x)$, it does not necessarily converge almost everywhere to $f(x)$. See Problem 7.16.

Solved Problems

L^p SPACES

7.1. Prove that $f(x) = 1/\sqrt[3]{x}$ (a) belongs to $L[0,8]$ but (b) does not belong to $L^3[0,8]$.

(a) The fact that $f(x)$ belongs to $L[0,8]$ follows from Problem 5.4, page 78.

(b) We must investigate the existence of

$$\int_0^8 \left| \frac{1}{\sqrt[3]{x}} \right|^3 dx = \int_0^8 \frac{dx}{x}$$

Since $1/x$ is unbounded in $[0,8]$, we proceed as in Chapter 5. Define

$$[1/x]_p = \begin{cases} 1/x & \text{for } 1/x \leq p \text{ or } x \geq 1/p \\ p & \text{for } 1/x > p \text{ or } x < 1/p \end{cases}$$

Then

$$\int_0^8 \frac{dx}{x} = \lim_{p \to \infty} \int_0^8 [1/x]_p \, dx$$

$$= \lim_{p \to \infty} \left[\int_0^{1/p} p \, dx + \int_{1/p}^8 \frac{dx}{x} \right]$$

$$= \lim_{p \to \infty} \left[px \Big|_0^{1/p} + \ln x \Big|_{1/p}^8 \right]$$

$$= \lim_{p \to \infty} [1 + \ln 8 + \ln p]$$

which does not exist. Thus $f(x) = 1/\sqrt[3]{x}$ does not belong to $L^3[0,8]$.

7.2. If $f(x) \in L^2$ and $g(x) \in L^2$, prove that $f(x)\,g(x) \in L$.

We have $\qquad\qquad (|f(x)| - |g(x)|)^2 \geq 0 \qquad\qquad\qquad (1)$

so that $\qquad |f(x)|^2 - 2|f(x)|\,|g(x)| + |g(x)|^2 \geq 0$

or $\qquad\qquad |f(x)|\,|g(x)| \leq \tfrac{1}{2}(|f(x)|^2 + |g(x)|^2)$

i.e. $\qquad\qquad |f(x)\,g(x)| \leq \tfrac{1}{2}(|f(x)|^2 + |g(x)|^2) \qquad\qquad (2)$

Then $\qquad \int_a^b |f(x)\,g(x)| \, dx \leq \tfrac{1}{2} \int_a^b |f(x)|^2 \, dx + \tfrac{1}{2} \int_a^b |g(x)|^2 \, dx \qquad (3)$

Since $f(x) \in L^2$ and $g(x) \in L^2$, the two integrals on the right of (3) exist so that the integral on the left of (3) exists, i.e. $f(x)\,g(x) \in L$.

Note that the result is a special case of Theorem 7-3 page 116, where $p = 2$.

7.3. If $f(x) \in L^2$, prove that $f(x) \in L$, i.e. $L^2 \subset L$.

This follows at once from Problem 7.2 by noting that $g(x) = 1 \in L^2$. Note, however, that the result is not valid if the interval $[a, b]$ is infinite.

INEQUALITIES

7.4. Prove *Schwarz's inequality*

$$\left| \int_a^b f(x)\, g(x)\, dx \right| \;\leq\; \left\{ \int_a^b |f(x)|^2\, dx \right\}^{1/2} \left\{ \int_a^b |g(x)|^2\, dx \right\}^{1/2}$$

where $f(x) \in L^2$, $g(x) \in L^2$.

For all real numbers λ we have, using the fact that $f(x) \in L^2$, $g(x) \in L^2$ and $f(x)\, g(x) \in L$,

$$\int_a^b [\lambda f(x) + g(x)]^2\, dx \;\geq\; 0$$

i.e. $A\lambda^2 + 2B\lambda + C \;\geq\; 0$ (1)

where $A = \int_a^b |f(x)|^2\, dx, \quad B = \int_a^b f(x)\, g(x)\, dx, \quad C = \int_a^b |g(x)|^2\, dx$ (2)

Now since $A > 0$, (1) can be written

$$\lambda^2 + \frac{2B}{A}\lambda + \frac{C}{A} \;\geq\; 0 \quad \text{or} \quad \left(\lambda + \frac{B}{A} \right)^2 + \frac{AC - B^2}{A^2} \;\geq\; 0$$

But this last inequality can be true for all real λ if and only if $AC - B^2 \geq 0$, i.e. $B^2 \leq AC$. Thus using (2) we find

$$\left\{ \int_a^b f(x)\, g(x)\, dx \right\}^2 \;\leq\; \left\{ \int_a^b |f(x)|^2\, dx \right\}\left\{ \int_a^b |g(x)|^2\, dx \right\}$$

from which we obtain as required,

$$\left| \int_a^b f(x)\, g(x)\, dx \right| \;\leq\; \left\{ \int_a^b |f(x)|^2\, dx \right\}^{1/2} \left\{ \int_a^b |g(x)|^2\, dx \right\}^{1/2}$$

7.5. Prove that if $f(x) \in L^2$, $g(x) \in L^2$,

$$\int_a^b |f(x)\, g(x)|\, dx \;\leq\; \left\{ \int_a^b |f(x)|^2\, dx \right\}^{1/2} \left\{ \int_a^b |g(x)|^2\, dx \right\}^{1/2}$$

This follows at once from Problem 7.4 on replacing $f(x)$ by $|f(x)|$ and $g(x)$ by $|g(x)|$.

7.6. (a) If u and v are non-negative real numbers, then their arithmetic mean is defined as $\frac{1}{2}(u + v)$ while their geometric mean is defined as the $\sqrt{uv}$. Prove that the geometric mean is less than or equal to the arithmetic mean.

(b) If we consider n_1 non-negative numbers all equal to u and n_2 non-negative numbers all equal to v, show that the geometric mean will be less than or equal to the arithmetic mean if and only if

$$(u^{n_1}\, v^{n_2})^{1/(n_1 + n_2)} \;\leq\; \frac{n_1 u + n_2 v}{n_1 + n_2}$$

(c) Show that the inequality in (b) is equivalent to

$$u^\alpha v^\beta \;\leq\; \alpha u + \beta v$$

where α, β are such that $\alpha + \beta = 1$.

(a) We must show that

$$\sqrt{uv} \;\leqq\; \tfrac{1}{2}(u+v) \quad\text{or}\quad u + v - 2\sqrt{uv} \;\geqq\; 0$$

i.e. $$(\sqrt{u} - \sqrt{v})^2 \;\geqq\; 0$$

But since this last statement is clearly true, we can reverse the steps to obtain the required inequality.

(b) The geometric mean of n positive numbers $x_1, x_2, \ldots x_n$ is defined as

$$\sqrt[n]{x_1 x_2 \ldots x_n} \;=\; (x_1 x_2 \ldots x_n)^{1/n}$$

while the arithmetic mean is defined as

$$\frac{x_1 + x_2 + \cdots + x_n}{n}$$

Then the geometric mean of n_1 numbers equal to u and n_2 numbers equal to v is

$$(u^{n_1} v^{n_2})^{1/(n_1 + n_2)}$$

while their arithmetic mean is

$$\frac{n_1 u + n_2 v}{n_1 + n_2}$$

so that the geometric mean is less than or equal to the arithmetic mean if and only if

$$(u^{n_1} v^{n_2})^{1/(n_1 + n_2)} \;\leqq\; \frac{n_1 u + n_2 v}{n_1 + n_2}$$

(c) The inequality in (b) can be written

$$u^{n_1/(n_1+n_2)} \; v^{n_2/(n_1+n_2)} \;\leqq\; \frac{n_1}{n_1 + n_2}\, u \;+\; \frac{n_2}{n_1 + n_2}\, v$$

Then letting $\alpha = \dfrac{n_1}{n_1 + n_2}$, $\beta = \dfrac{n_2}{n_1 + n_2}$ so that $\alpha + \beta = 1$ we obtain as required,

$$u^{\alpha} v^{\beta} \;\leqq\; \alpha u + \beta v$$

The validity of the inequality in (b) showing that the geometric mean is actually less than or equal to the arithmetic mean, is shown in Problem 7.7.

7.7. Prove that for any non-negative real numbers u, v and positive numbers α, β such that $\alpha + \beta = 1$,
$$u^{\alpha} v^{\beta} \;\leqq\; \alpha u + \beta v$$

thus demonstrating the truth of the inequality in Problem 7.6(b).

If $u = 0$ the inequality is trivial so that we can assume $u > 0$. Then letting $v = ux$ and $\alpha + \beta = 1$, the required inequality can be written

$$1 + \beta x - x^{\beta} - \beta \;\geqq\; 0 \tag{1}$$

Denoting the left side by $F(x)$, it follows that the derivative is $F'(x) = \beta(1 - x^{\beta-1})$.

Now $F(x) = 0$ for $x = 1$. Thus since $F'(x) < 0$ for $0 < x < 1$, it follows that $F(x) \geqq 0$ for $0 < x < 1$, i.e. the inequality (1) is true for $0 < x < 1$.

Similarly since $F'(x) > 0$ for $x > 1$, it follows that $F(x) > 0$ for $x > 1$, i.e. the inequality (1) is true for $x > 1$. Also equality holds for $x = 1$.

Thus (1) is true for all $x > 0$ and so the required inequality is proved.

7.8.   Prove Holder's inequality

$$\left| \int_a^b f(x)\, g(x)\, dx \right| \;\leq\; \left\{ \int_a^b |f(x)|^p\, dx \right\}^{1/p} \left\{ \int_a^b |g(x)|^q\, dx \right\}^{1/q}$$

where $p > 0$, $q > 0$, $1/p + 1/q = 1$ and $f(x) \in L^p$, $g(x) \in L^q$.

In the inequality of Problem 7.7 let

$$u = \frac{|f(x)|^p}{F}, \quad v = \frac{|g(x)|^q}{G}, \quad \alpha = \frac{1}{p}, \quad \beta = \frac{1}{q}$$

where

$$F = \int_a^b |f(x)|^p\, dx, \qquad G = \int_a^b |g(x)|^q\, dx \tag{1}$$

Then

$$\left\{ \frac{|f(x)|^p}{F} \right\}^{1/p} \left\{ \frac{|g(x)|^q}{G} \right\}^{1/q} \;\leq\; \frac{|f(x)|^p}{pF} + \frac{|g(x)|^q}{qG} \tag{2}$$

Thus by integrating from a to b and using the definition (1) of F and G, we have from (2)

$$\int_a^b \left\{ \frac{|f(x)|^p}{F} \right\}^{1/p} \left\{ \frac{|g(x)|^q}{G} \right\}^{1/q} dx \;\leq\; \frac{1}{p} + \frac{1}{q} = 1 \tag{3}$$

Since F and G are independent of x, we can multiply both sides of (3) by $F^{1/p} G^{1/q}$ to obtain

$$\int_a^b |f(x)|\, |g(x)|\, dx \;\leq\; \left\{ \int_a^b |f(x)|^p\, dx \right\}^{1/p} \left\{ \int_a^b |g(x)|^q\, dx \right\}^{1/q}$$

The required inequality follows on noting that

$$\left| \int_a^b f(x)\, g(x)\, dx \right| \;\leq\; \int_a^b |f(x)|\, |g(x)|\, dx$$

7.9.   Prove *Minkowski's inequality*

$$\left\{ \int_a^b |f(x) + g(x)|^p\, dx \right\}^{1/p} \;\leq\; \left\{ \int_a^b |f(x)|^p\, dx \right\}^{1/p} + \left\{ \int_a^b |g(x)|^p\, dx \right\}^{1/p}$$

where $p \geq 1$.

If $p = 1$, the inequality is trivial. For $p > 1$, we have using Holder's inequality

$$\int_a^b |f + g|^p\, dx \;=\; \int_a^b |f + g|\, |f + g|^{p-1}\, dx$$

$$\leq\; \int_a^b |f|\, |f + g|^{p-1}\, dx \;+\; \int_a^b |g|\, |f + g|^{p-1}\, dx$$

$$\leq\; \left\{ \int_a^b |f|^p\, dx \right\}^{1/p} \left\{ \int_a^b |f + g|^p\, dx \right\}^{(p-1)/p}$$

$$+ \left\{ \int_a^b |g|^p\, dx \right\}^{1/p} \left\{ \int_a^b |f + g|^p\, dx \right\}^{(p-1)/p}$$

Then dividing both sides by $\left\{ \int_a^b |f + g|^p\, dx \right\}^{(p-1)/p}$ [which is assumed to be different from zero

since otherwise the inequality is trivial], the required result follows.

THEOREMS INVOLVING FUNCTIONS IN L^p SPACES

7.10.   Prove Theorem 7-2, page 116: If $g(x) \in L^p$ and $|f(x)| \leq |g(x)|$, then $f(x) \in L^p$.

If $|f(x)| \leq |g(x)|$, then $|f(x)|^p \leq |g(x)|^p$. Thus

$$\int_a^b |f(x)|^p\, dx \;\leq\; \int_a^b |g(x)|^p\, dx$$

so that if the right side exists, i.e. if $g(x) \in L^p$, then the left side also exists, i.e. $f(x) \in L^p$.

7.11. Prove Theorem 7-3, page 116: If $f(x) \in L^p$ and $g(x) \in L^p$, then $f(x)\, g(x) \in L^{p/2}$.

Since $f(x) \in L^p$ and $g(x) \in L^p$, we have

$$\int_a^b |f(x)|^p\, dx \;<\; \infty, \qquad \int_a^b |g(x)|^p\, dx \;<\; \infty$$

i.e. the integrals exist. Then by Schwarz's inequality on replacing $f(x)$ by $|f(x)|^{p/2}$ and $g(x)$ by $|g(x)|^{p/2}$, we have

$$\int_a^b |f(x)\, g(x)|^{p/2}\, dx \;\leqq\; \left\{ \int_a^b |f(x)|^p\, dx \right\}^{1/2} \left\{ \int_a^b |g(x)|^p\, dx \right\}^{1/2}$$

Since the integrals on the right exist, the integral on the left exists, i.e. $f(x)\, g(x) \in L^{p/2}$.

THE L^p SPACES AS METRIC SPACES

7.12. Prove that an L^p space is a metric space.

The distance between two elements (functions) of L^p is defined as

$$D(f, g) \;=\; \left\{ \int_a^b |f(x) - g(x)|^p\, dx \right\}^{1/p}$$

In order to prove that L^p is a metric space, we must satisfy the requirements 1-4 on page 6, i.e.

1. $D(f, g) \geqq 0$

2. $D(f, g) = D(g, f)$

3. $D(f, g) = 0$ if and only if $f = g$

4. $D(f, g) \leqq D(f, h) + D(h, g)$

Now requirements 1 and 2 are obviously satisfied from the definition of $D(f, g)$. Requirement 3 is satisfied if we consider the distance between f and g as zero, i.e. if and only if the functions have the same values almost everywhere and we shall henceforth use this interpretation.

Finally requirement 4 follows as a consequence of Minkowski's inequality.

MEAN CONVERGENCE, CAUCHY SEQUENCES AND THE RIESZ-FISCHER THEOREM

7.13. Prove that if a sequence $\langle f_n(x) \rangle$ converges in the mean to two different functions $f(x)$ and $g(x)$, then $f(x) = g(x)$ almost everywhere. Thus if l.i.m. $f_n(x)$ exists, it is unique.
 $n \to \infty$

If $\langle f_n(x) \rangle$ converges in mean to $f(x)$, then

$$\lim_{n \to \infty} \int_a^b |f_n(x) - f(x)|^p\, dx \;=\; 0 \qquad\qquad (1)$$

Similarly if $\langle f_n(x) \rangle$ converges in mean to $g(x)$, then

$$\lim_{n \to \infty} \int_a^b |f_n(x) - g(x)|^p\, dx \;=\; 0 \qquad\qquad (2)$$

But by Minkowski's inequality,

$$\left\{ \int_a^b |f(x) - g(x)|^p\, dx \right\}^{1/p} \;\leqq\; \left\{ \int_a^b |f(x) - f_n(x)|^p\, dx \right\}^{1/p} + \left\{ \int_a^b |f_n(x) - g(x)^p\, dx \right\}^{1/p}$$

Thus letting $n \to \infty$ and using (1) and (2), we have

$$\int_a^b |f(x) - g(x)|^p\, dx \;=\; 0$$

or $f(x) = g(x)$ almost everywhere.

7.14. Prove Theorem 7-7, page 117: If a sequence $\langle f_n(x) \rangle$ converges in the mean to $f(x)$ in the space L^p, then it is a Cauchy sequence.

By Minkowski's inequality we have

$$\left\{ \int_a^b |f_m(x) - f_n(x)|^p\, dx \right\}^{1/p} \;\leqq\; \left\{ \int_a^b |f_m(x) - f(x)|^p\, dx \right\}^{1/p}$$
$$+ \left\{ \int_a^b |f(x) - f_n(x)|^p\, dx \right\}^{1/p}$$

or $\|f_m - f_n\| \;\leqq\; \|f_m - f\| + \|f_n - f\|$

But since $\langle f_m(x) \rangle$ and $\langle f_n(x) \rangle$ both converge in the mean to $f(x)$, it follows that given $\epsilon > 0$ there exists $n_0 > 0$ such that for $m > n_0$, $n > n_0$,

$$\|f_m - f\| \;<\; \frac{\epsilon}{2}, \quad \|f_n - f\| \;<\; \frac{\epsilon}{2}$$

Then $\|f_m - f_n\| \;<\; \dfrac{\epsilon}{2} + \dfrac{\epsilon}{2} \;=\; \epsilon \quad$ for $\quad m > n_0,\ n > n_0$

and the required result follows.

7.15. If $\operatorname*{l.i.m.}_{n \to \infty} f_n(x) = f(x)$, prove that

$$\lim_{n \to \infty} \int_a^b |f_n(x)|^p\, dx \;=\; \int_a^b |f(x)|^p\, dx$$

or equivalently, $\lim_{n \to \infty} \|f_n\| = \|f\|$.

We have from Minkowski's inequality,

$$\|f_n\| \;=\; \|(f_n - f) + f\| \;\leqq\; \|f_n - f\| + \|f\|$$

or taking the limit as $n \to \infty$,

$$\lim_{n \to \infty} \|f_n\| \;\leqq\; \|f\| \tag{1}$$

Similarly $\|f\| \;=\; \|f - f_n + f_n\| \;\leqq\; \|f - f_n\| + \|f_n\|$

or taking the limit as $n \to \infty$,

$$\|f\| \;\leqq\; \lim_{n \to \infty} \|f_n\| \tag{2}$$

Then from (1) and (2) we have

$$\lim_{n \to \infty} \|f_n\| \;=\; \|f\|$$

The result is often interpreted by saying that the norm of the metric space L^p is continuous.

7.16. Let

$$f_n(x) \;=\; \begin{cases} n, & 0 < x < 1/n \\ 0, & 1/n < x < 1 \end{cases} \qquad n = 1, 2, \ldots$$

Show that in the space L^2 (a) $\lim_{n \to \infty} f_n(x) = 0$ while (b) $\operatorname*{l.i.m.}_{n \to \infty} f_n(x) \neq 0$, and (c) interpret the results.

(a) Let x_0 be any point of the interval $0 < x < 1$. Since $1/n < x_0 < 1$ for $n > 1/x_0 = n_0$, we have $|f_n(x_0) - 0| = 0 < \epsilon$ for $n > n_0$, i.e. $\lim_{n \to \infty} f_n(x) = 0$ for all x_0 in $[0, 1]$.

(b) If $\operatorname*{l.i.m.}_{n \to \infty} f_n(x) = 0$, then we would have

$$\lim_{n \to \infty} \int_0^1 |f_n(x)|^2\, dx \;=\; 0$$

However, $$\int_0^1 |f_n(x)|^2 \, dx \;=\; \int_0^{1/n} n^2 \, dx \;+\; \int_{1/n}^1 0^2 \, dx \;=\; n$$

and so $$\lim_{n \to \infty} \int_0^1 |f_n(x)|^2 \, dx \;=\; \infty, \quad \text{i.e.} \quad \underset{n \to \infty}{\text{l.i.m.}} \, f_n(x) \;\neq\; 0$$

(c) The result shows that a sequence $\langle f_n(x) \rangle$ may converge to $f(x)$ at all points of an interval without converging in the mean to $f(x)$ on the interval. The example can be extended to the case L^p where $p > 1$ [see Problem 7.51].

Conversely we can show that a sequence $\langle f_n(x) \rangle$ may converge in the mean to $f(x)$ on an interval without converging to $f(x)$ in the usual (pointwise) sense [see Problem 7.54].

7.17. Prove the Riesz-Fischer theorem [Theorem 7-8, page 117].

Since the sequence of functions $\langle f_n(x) \rangle$ in L^p is a Cauchy sequence, i.e.

$$\lim_{\substack{m \to \infty \\ n \to \infty}} \int_a^b |f_m(x) - f_n(x)|^p \, dx \;=\; 0$$

it follows that to each natural number ν there is a smallest natural number n_ν such that

$$\int_a^b |f_m(x) - f_n(x)|^p \, dx \;<\; \frac{1}{3^\nu} \quad \text{for} \quad m \geqq n_\nu, \; n \geqq n_\nu$$

We can choose in particular $m = n_{\nu+1}, \; n = n_\nu$ so that

$$\int_a^b |f_{n_{\nu+1}}(x) - f_{n_\nu}(x)|^p \, dx \;<\; \frac{1}{3^\nu} \quad \text{for} \quad \nu = 1, 2, \ldots \tag{1}$$

Let us define $$E_\nu \;=\; \left\{ x : \; |f_{n_{\nu+1}}(x) - f_{n_\nu}(x)| > \frac{1}{2^{\nu/p}} \right\} \tag{2}$$

Then $$\int_{E_\nu} |f_{n_{\nu+1}}(x) - f_{n_\nu}(x)|^p \, dx \;\geqq\; \int_{E_\nu} \left(\frac{1}{2^{\nu/p}} \right)^p dx \;=\; \frac{1}{2^\nu} m(E_\nu) \tag{3}$$

Combining this with (1),
$$\frac{1}{2^\nu} m(E_\nu) \;<\; \frac{1}{3^\nu} \quad \text{or} \quad m(E_\nu) \;<\; \left(\tfrac{2}{3} \right)^\nu$$

Consider now

$$E_N \cup E_{N+1} \cup \cdots \;=\; \left\{ x : \; |f_{n_{N+1}} - f_{n_N}| > \frac{1}{2^{N/p}}, \; |f_{n_{N+2}} - f_{n_{N+1}}| > \frac{1}{2^{(N+1)/p}}, \; \cdots \right\} \tag{4}$$

If x does not belong to this set, it is clear that

$$|f_{n_{N+1}} - f_{n_N}| \;\leqq\; \frac{1}{2^{N/p}}, \quad |f_{n_{N+2}} - f_{n_{N+1}}| \;\leqq\; \frac{1}{2^{(N+1)/p}}, \quad \cdots$$

so that the series $$\sum_{\nu=1}^\infty |f_{n_{\nu+1}} - f_{n_\nu}| \tag{5}$$
converges.

But the measure of the set (4) is less than or equal to

$$m(E_N) + m(E_{N+1}) + \cdots \;\leqq\; \left(\tfrac{2}{3} \right)^N + \left(\tfrac{2}{3} \right)^{N+1} + \cdots \;=\; 3 \left(\tfrac{2}{3} \right)^N$$

which approaches zero as $N \to \infty$.

This shows that the set of all x for which (5) does not converge is of measure zero or, in other words, the series (5) converges almost everywhere.

Now since the series (5) converges almost everywhere, so also does the series

$$\sum_{\nu=1}^{\infty} (f_{n_{\nu+1}} - f_{n_\nu})$$

since an absolutely convergent series is convergent. Writing

$$f_{n_k} = f_{n_1} + \sum_{\nu=1}^{k-1} (f_{n_{\nu+1}} - f_{n_\nu})$$

it follows that

$$\lim_{k \to \infty} f_{n_k} = f_{n_1} + \sum_{\nu=1}^{\infty} (f_{n_{\nu+1}} - f_{n_\nu})$$

exists. Denoting this limit by $f(x)$ we thus see that

$$\lim_{k \to \infty} f_{n_k}(x) = f(x) \quad \text{almost everywhere}$$

We must now show that $\langle f_n(x) \rangle$ does in fact converge in mean to $f(x)$. We first prove that $f_{n_k}(x)$ converges in mean to $f(x)$. To do this we observe that by Fatou's theorem,

$$\varliminf_{\nu \to \infty} \int_a^b |f_{n_k}(x) - f_{n_\nu}(x)|^p \, dx \geqq \int_a^b |f_{n_k}(x) - f(x)|^p \, dx \tag{6}$$

But from the fact that $\langle f_n(x) \rangle$ is a Cauchy sequence we see that for every $\epsilon > 0$ there is a number K such that

$$\int_a^b |f_{n_k}(x) - f_{n_\nu}(x)|^p \, dx < \epsilon \quad \text{for} \quad k > K, \ \nu > K$$

Thus from (6),
$$\int_a^b |f_{n_k}(x) - f(x)|^p \, dx \leqq \epsilon \quad \text{for} \quad k > K$$

which shows that $f_{n_k}(x)$ converges in mean to $f(x)$ or equivalently

$$\lim_{n_k \to \infty} \int_a^b |f_{n_k}(x) - f(x)|^p \, dx = 0$$

To show that $f_n(x)$ converges in mean to $f(x)$ we note that by Minkowski's inequality

$$\left\{ \int_a^b |f_n(x) - f(x)|^p \, dx \right\}^{1/p} \leqq \left\{ \int_a^b |f_n(x) - f_{n_k}(x)|^p \, dx \right\}^{1/p} + \left\{ \int_a^b |f_{n_k}(x) - f(x)|^p \, dx \right\}^{1/p}$$

so that
$$\lim_{n \to \infty} \int_a^b |f_n(x) - f(x)|^p \, dx = 0$$

i.e. $f_n(x)$ converges in mean to $f(x)$.

The limit function $f(x)$ is unique apart from a set of measure zero or, in other words, if there are two limit functions $f(x)$ and $g(x)$ then $f(x) = g(x)$ almost everywhere.

CONVERGENCE IN MEASURE

7.18. Prove Theorem 7-10, page 117: If $\langle f_n(x) \rangle$ converges in the mean to $f(x)$, then it converges in measure to $f(x)$.

Let $\sigma > 0$ and $E_n = \{x : |f_n - f| \geqq \sigma\}$. Then

$$\int_{E_n} |f_n(x) - f(x)|^p \, dx \geqq \int_{E_n} \sigma^p \, dx = \sigma^p \, m(E_n)$$

Since the left side approaches zero as $n \to \infty$, we see that $\lim_{n \to \infty} m(E_n) = 0$, i.e. $\langle f_n(x) \rangle$ converges in measure to $f(x)$.

7.19. Prove Theorem 7-11, page 117: If $\langle f_n(x) \rangle$ converges in measure to $f(x)$ on a set E, then there is a subsequence $f_{n_1}(x), f_{n_2}(x), \ldots, n_1 < n_2 < \cdots$, i.e. $\langle f_{n_k}(x) \rangle$, which converges almost everywhere to $f(x)$.

Consider a sequence of positive numbers $\sigma_1 > \sigma_2 > \sigma_3 > \cdots$ such that $\lim\limits_{k \to \infty} \sigma_k = 0$ and let $\kappa_1 + \kappa_2 + \cdots$ be a convergent series in which each term is positive.

Choose the natural number n_1 such that for all $x \in E$,

$$m\{x : |f_{n_1} - f| \geqq \sigma_1\} < \kappa_1$$

This number must exist because by hypothesis

$$\lim_{n \to \infty} m\{x : |f_n - f| \geqq \sigma_1\} = 0$$

Similarly let $n_2 > n_1$ be a natural number such that

$$m\{x : |f_{n_2} - f| \geqq \sigma_2\} < \kappa_2$$

and in general $n_k > n_{k-1} > \cdots > n_1$ such that

$$m\{x : |f_{n_k} - f| \geqq \sigma_k\} < \kappa_k$$

This procedure defines the sequence $\langle n_k \rangle$.

We now prove that $\lim\limits_{k \to \infty} f_{n_k}(x) = f(x)$. To do this suppose that

$$P_j = \bigcup_{k=j}^{\infty} \{x : |f_{n_k} - f| \geqq \sigma_k\}, \qquad Q = \bigcap_{j=1}^{\infty} P_j$$

Then since $P_1 \supset P_2 \supset P_3 \supset \cdots$, it follows from Theorem 2-15, page 33, that

$$m(Q) = \lim_{j \to \infty} m(P_j) \tag{1}$$

Now since $m(P_j) \leqq \sum\limits_{k=j}^{\infty} \kappa_k$ it follows that $\lim\limits_{j \to \infty} m(P_j) = 0$ and so from (1)

$$m(Q) = 0$$

If we can now show that

$$\lim_{k \to \infty} f_{n_k}(x) = f(x)$$

for all $x \in E - Q$, the required result will follow. To do this suppose that $x_0 \in E - Q$. Then there is some natural number j_0 such that $x_0 \notin P_{j_0}$, so that from the definition of P_j

$$x_0 \notin \{x : |f_{n_k} - f| \geqq \sigma_k\}$$

for all $k \geqq j_0$. It follows that

$$|f_{n_k}(x_0) - f(x_0)| < \sigma_k \tag{2}$$

But since $\sigma_k \to 0$ as $k \to \infty$, we see from (2) that

$$\lim_{k \to \infty} f_{n_k}(x_0) = f(x_0)$$

i.e. for all $x \in E - Q$

$$\lim_{k \to \infty} f_{n_k}(x) = f(x)$$

Supplementary Problems

L^p SPACES

7.20. Prove that the function $f(x)$ of Problem 7.1 (a) belongs to $L^p[0, 8]$ if $1 \leqq p < 3$ but (b) does not belong to $L^p[0, 8]$ if $p \geqq 3$. (c) For what values of p does $f(x) = 1/\sqrt{x}$ belong to $L^p[0, 1]$?

7.21. Prove that if $f(x) \in L^p$, $p \geqq 1$, then $c\,f(x) \in L^p$ for any constant c.

7.22. If $f(x) \in L^2$ and $g(x) \in L^2$, prove that (a) $[f(x) + g(x)] \in L^2$, (b) $[f(x) - g(x)] \in L^2$.

7.23. If $f_1(x), f_2(x), \ldots, f_n(x)$ all belong to L^2 and $c_1, c_2, \ldots, c_n$ are any constants, prove that $c_1 f_1(x) + c_2 f_2(x) + \cdots + c_n f_n(x) \in L^2$.

7.24. If a function belongs to L, does it also belong to L^2? Explain.

INEQUALITIES AND THEOREMS INVOLVING FUNCTIONS IN L^p SPACES

7.25. Prove that the equality holds in Schwarz's inequality if and only if $f(x)/g(x)$ is constant almost everywhere.

7.26. Prove the following analog of Schwarz's inequality for real numbers a_k, b_k, $k = 1, \ldots, n$,

$$|a_1 b_1 + \cdots + a_n b_n|^2 \;\leqq\; (|a_1|^2 + \cdots + |a_n|^2)(|b_1|^2 + \cdots + |b_n|^2)$$

and show that the equality holds if and only if a_k/b_k is constant.

7.27. Obtain Schwarz's inequality for integrals by using Problem 7.26 and the definition of an integral as a limit of a sum.

7.28. Prove that if $f(x) \in L^2$, then

$$\int_a^b |f(x)|\, dx \;\leqq\; \sqrt{b-a} \left[\int_a^b |f(x)|^2\, dx \right]^{1/2}$$

Thus show that $L^2 \subset L$.

7.29. Prove that $f(x) = e^{-x}/\sqrt{x} \in L[0, \infty]$.

7.30. Prove Minkowski's inequality for $p = 2$ by direct use of Schwarz's inequality.

7.31. Prove that the equality in Minkowski's inequality holds if and only if $f(x)/g(x)$ is constant almost everywhere.

7.32. If a_k, b_k, $k = 1, \ldots, n$, are real numbers and $p > 1$, prove that

$$\{|a_1 + b_1|^p + \cdots + |a_n + b_n|^p\}^{1/p} \;\leqq\; \{|a_1|^p + \cdots + |a_n|^p\}^{1/p} + \{|b_1|^p + \cdots + |b_n|^p\}^{1/p}$$

where the equality holds if and only if a_k/b_k is constant. Discuss the relationship with Minkowski's inequality.

7.33. Prove that $\|f + g\| \leqq \|f\| + \|g\|$ and interpret geometrically.

7.34. Discuss the significance of $\|f\|_p$ in case $p = 1$. In this case what is $\|f - g\|_p$?

7.35. Is L^∞ a metric space? Explain.

7.36. Prove that if $f(x) \in L^p,\ p > 1,$ then

$$\int_a^b |f(x)|\, dx \;\leq\; (b-a)^{p/(p-1)} \left[\int_a^b |f(x)|^p\, dx \right]^{1/p}$$

Thus show that $L^p \subset L$ for $p \geqq 1$, generalizing the result of Problem 7.3 and proving the first part of Theorem 7-1, page 116.

7.37. Prove the second part of Theorem 7-1, page 116, i.e. $L^p \subset L^n$ if $p > n \geqq 1$.

7.38. Discuss Holder's inequality in case $p = 1$.

7.39. Prove that

$$\left| \int_0^{2\pi} \frac{\cos x}{\sqrt{x^2+1}}\, dx \right| \;\leqq\; \pi \tan^{-1} 2\pi$$

7.40. Prove that the equality in Holder's inequality holds if and only if $f(x)/g(x)$ is constant almost everywhere.

7.41. Prove Theorem 7-4, page 116.

7.42. Prove Theorem 7-5, page 116.

7.43. Prove that if $f_1(x) \in L^p$ and $f_2(x) \in L^p$, then $c_1 f_1(x) + c_2 f_2(x) \in L^p$ where c_1 and c_2 are any constants.

7.44. Generalize the result of Problem 7.43 to n functions $f_1(x), \ldots, f_n(x)$.

7.45. State and prove an inequality for real numbers corresponding to Holder's inequality.

7.46. Prove that the geometric mean of any set of non-negative numbers is less than or equal to their arithmetic mean. Is there a corresponding result involving integrals?

MEAN CONVERGENCE, CAUCHY SEQUENCES AND THE RIESZ-FISCHER THEOREM

7.47. If $\langle f_n(x) \rangle$ converges in mean to $f(x)$, prove that $\|f_n\|$ is bounded.

7.48. If $\underset{n \to \infty}{\text{l.i.m.}}\, f_n(x) = f(x)$ and $\underset{n \to \infty}{\text{l.i.m.}}\, g_n(x) = g(x)$ where all functions are in L^p, is it true that
(a) $\underset{n \to \infty}{\text{l.i.m.}}\, [f_n(x) + g_n(x)] = f(x) + g(x)$ (b) $\underset{n \to \infty}{\text{l.i.m.}}\, f_n(x)\, g_n(x) = f(x)\, g(x)$ (c) $\underset{n \to \infty}{\lim}\, \|f_n(x) + g_n(x)\| = \|f(x) + g(x)\|$? Explain.

7.49. Explain the relationship between Theorems 7-7 and 7-8, page 117.

7.50. Prove that a sequence of functions in Hilbert space converges in the mean to a function in Hilbert space if and only if the sequence is a Cauchy sequence.

7.51. Obtain a generalization of the remarks of Problem 7.16 for L^p space, $p > 1$.

7.52. Prove that if $\langle f_n(x) \rangle$ converges in mean to $f(x)$ in L^p and $g(x) \in L^q$, then

$$\lim_{n \to \infty} \int_a^b f_n(x)\, g(x)\, dx \;=\; \int_a^b f(x)\, g(x)\, dx$$

7.53. A sequence $\langle f_n(x) \rangle$ in L^p is said to *converge weakly* to $f(x)$ in L^p if for every function $w(x) \in L^q$, where $1/p + 1/q = 1$,

$$\lim_{n \to \infty} \int_a^b w(x)\, f_n(x)\, dx \;=\; \int_a^b w(x)\, f(x)\, dx$$

Prove that if $\langle f_n(x) \rangle$ converges in the mean to $f(x)$ in L^p, then it also converges weakly to $f(x)$.

7.54. Give an example to illustrate that if $\langle f_n(x) \rangle$ converges in the mean to $f(x)$, it does not necessarily converge almost everywhere to $f(x)$.

7.55. Under what conditions will $\sum_{k=1}^{\infty} c_k f_k(x) \in L^p$ if $f_k(x) \in L^p$, $k = 1, 2, \ldots$, and c_k are constants? Justify your conclusions.

7.56. Let $f_n(x) = \dfrac{n}{1 + n\sqrt{x}}$ for $0 \leqq x \leqq 1$ where $n = 1, 2, \ldots$. (a) Is $\langle f_n(x) \rangle$ a Cauchy sequence? (b) Does $f_n(x) \in L^2$ for $n = 1, 2, \ldots$? (c) Does $\lim\limits_{n \to \infty} f_n(x) \in L^2$? (d) Does $\underset{n \to \infty}{\text{l.i.m.}} f_n(x) \in L^2$? Discuss your results in connection with the Riesz-Fischer theorem.

CONVERGENCE IN MEASURE

7.57. Prove that if a sequence of functions $\langle f_n(x) \rangle$ converges in measure to two different functions $f(x)$ and $g(x)$, then $f(x) = g(x)$ almost everywhere. Can you say that convergence in measure is unique? Explain.

7.58. Prove Fatou's theorem [see page 75] if the sequence $\langle f_n(x) \rangle$ converges in measure to $f(x)$.

7.59. Suppose that $\langle f_n(x) \rangle$ converges in measure to $f(x)$ in Hilbert space. Prove that if $\|f_n\|$ is uniformly bounded, then $\langle f_n(x) \rangle$ converges weakly to $f(x)$ [see Problem 7.53].

7.60. Is the result of Problem 7.59 true for L^p spaces in general? Explain.

7.61. Prove Theorem 7-9, page 117.

Chapter 8

Applications to Fourier Series

DEFINITION OF A FOURIER SERIES

Let $f(x)$ be Lebesgue integrable in $(-\pi, \pi)$. Suppose that $f(x)$ is defined to be periodic with period 2π outside of this interval, i.e. $f(x \pm 2k\pi) = f(x)$, $k = 1, 2, \ldots$. The *trigonometric series*

$$\frac{a_0}{2} + \sum_{n=1}^{\infty} (a_n \cos nx + b_n \sin nx) \tag{1}$$

is called the *Fourier series* corresponding to $f(x)$ if

$$a_n = \frac{1}{\pi} \int_{-\pi}^{\pi} f(x) \cos nx \, dx, \qquad b_n = \frac{1}{\pi} \int_{-\pi}^{\pi} f(x) \sin nx \, dx \tag{2}$$

We call a_n and b_n the *Fourier coefficients* corresponding to $f(x)$.

Since $f(x)$ is periodic with period 2π, any other interval of length 2π can be used instead of $(-\pi, \pi)$ such as for example $(0, 2\pi)$ or in general $(c, c + 2\pi)$ where c is any constant. In such case the integration limits $-\pi$ and π in (2) are changed to c and $c + 2\pi$ respectively.

Extension to the case where $f(x)$ has period $2l$ where $l > 0$, is easily made [see Problem 8.36].

THE RIEMANN-LEBESGUE THEOREM

If $f(x)$ is integrable in (a, b), then

$$\lim_{\alpha \to \infty} \int_a^b f(x) \cos \alpha x \, dx = 0, \qquad \lim_{\alpha \to \infty} \int_a^b f(x) \sin \alpha x \, dx = 0 \tag{3}$$

In particular the Fourier coefficients (2) approach zero as $n \to \infty$.

CONVERGENCE OF FOURIER SERIES

An important question which naturally arises is whether a Fourier series corresponding to an integrable function $f(x)$ will converge and, if it does, whether it will converge to $f(x)$. To answer this question we need to consider the partial sums of the series (1) given by

$$S_M(x) = \frac{a_0}{2} + \sum_{n=1}^{M} (a_n \cos nx + b_n \sin nx) \tag{4}$$

Using the coefficients (2) we can show [Problem 8.7] that

$$S_M(x) = \frac{1}{2\pi} \int_{-\pi}^{\pi} f(x+u) \frac{\sin (M + \frac{1}{2})u}{\sin \frac{1}{2}u} \, du \tag{5}$$

or [Problem 8.8]

$$S_M(x) = \frac{1}{2\pi} \int_0^{\pi} [f(x+t) + f(x-t)] \frac{\sin (M + \frac{1}{2})t}{\sin \frac{1}{2}t} \, dt \tag{6}$$

Subtracting $S(x)$ from both sides of (6), we then find [Problem 8.9] that

$$S_M(x) - S(x) = \frac{1}{2\pi} \int_0^\pi [f(x+t) + f(x-t) - 2S(x)] \frac{\sin(M + \frac{1}{2})t}{\sin \frac{1}{2}t} \, dt \qquad (7)$$

We thus arrive at the following important

Theorem 8-1.   The Fourier series converges to $S(x)$ if and only if the integral in (7) approaches zero as $M \to \infty$. In such case $\lim\limits_{M \to \infty} S_M(x) = S(x)$.

Other important related theorems are the following.

Theorem 8-2.   The Fourier series converges to $S(x)$ if and only if for some fixed number δ such that $0 < \delta \le \pi$,

$$\lim_{M \to \infty} \int_0^\delta [f(x+t) + f(x-t) - 2S(x)] \frac{\sin(M + \frac{1}{2})t}{\sin \frac{1}{2}t} \, dt = 0$$

Theorem 8-3.   The Fourier series converges to $S(x)$ if and only if for some fixed number δ such that $0 < \delta \le \pi$,

$$\lim_{M \to \infty} \int_0^\delta [f(x+t) + f(x-t) - 2S(x)] \frac{\sin(M + \frac{1}{2})t}{t} \, dt = 0$$

SUFFICIENT CONDITIONS FOR CONVERGENCE OF FOURIER SERIES

Various sufficient conditions exist under which a Fourier series will converge to $S(x)$. The following are two such conditions or *tests* as they are often called.

Theorem 8-4 [**Dini's condition**]. The Fourier series converges to $S(x)$ if for some fixed number δ such that $0 < \delta \le \pi$,

$$\int_0^\delta \frac{f(x+t) + f(x-t) - 2S(x)}{t} \, dt \text{ exists}$$

Theorem 8-5 [**Jordan's condition**]. The Fourier series converges to

$$S(x) = \frac{f(x+0) + f(x-0)}{2}$$

if $f(t)$ is of bounded variation in a neighborhood of $t = x$. If furthermore $f(t)$ is continuous at $t = x$, then the Fourier series converges to $f(x)$.

For other sufficient conditions see Problems 8.49, 8.50 and 8.55.

INTEGRATION OF FOURIER SERIES

Theorem 8-6.   Let the Fourier series corresponding to the integrable function $f(x)$ be given by

$$\frac{a_0}{2} + \sum_{n=1}^\infty (a_n \cos nx + b_n \sin nx) \qquad (8)$$

Then if we integrate the series term by term from α to β, the resulting series converges to $\int_\alpha^\beta f(x) \, dx$.

The remarkable thing about this theorem is that the series (8) need not be convergent.

For a generalization of this theorem see Problem 8.60.

FOURIER SERIES IN L^2 SPACES

Thus far we have been concerned with pointwise convergence of Fourier series. Two other important types of convergence can also be considered. The first involves mean convergence in L^2 spaces which is considered in this chapter and the second involves the concept of *summability* which is considered in Appendix B, page 175.

It turns out that the theory of mean convergence of Fourier series in L^2 spaces is closely related to the theory of orthogonal (or orthonormal) series, since a Fourier series is a special case of an orthogonal series. For this reason we shall treat this more general theory.

ORTHOGONAL FUNCTIONS

Two functions $f_1(x)$ and $f_2(x)$ are said to be *orthogonal* in (a, b) if

$$\int_a^b f_1(x) f_2(x)\, dx \;=\; 0 \tag{9}$$

A set of functions $f_1(x), f_2(x), \ldots$ is said to be an *orthogonal* set in (a, b) if

$$\int_a^b f_m(x) f_n(x)\, dx \;=\; 0 \qquad m \neq n \tag{10}$$

and the functions are said to be *mutually orthogonal* or simply *orthogonal* in (a, b).

Example: The functions $\{\sin mx\}$, $m = 1, 2, \ldots$, are mutually orthogonal in $(-\pi, \pi)$ since

$$\int_{-\pi}^{\pi} \sin mx \, \sin nx \, dx \;=\; 0 \qquad m \neq n$$

ORTHONORMAL FUNCTIONS

If the set of functions $f_1(x), f_2(x), \ldots$ is orthogonal in (a, b), then if $f_n(x) \in L^2$ we will have

$$\int_a^b [f_n(x)]^2\, dx \;=\; A_n \qquad n = 1, 2, \ldots$$

where $A_n > 0$ is finite. Then if we let $\phi_n(x) = f_n(x)/\sqrt{A_n}$, we have

$$\int_a^b [\phi_n(x)]^2\, dx \;=\; 1 \tag{11}$$

$$\int_a^b \phi_m(x)\, \phi_n(x)\, dx \;=\; 0 \qquad m \neq n \tag{12}$$

The set of functions $\{\phi_n(x)\}$ is said to be *orthogonal* and *normalized* in (a, b) and is referred to as an *orthonormal* set of functions in (a, b) and the functions are said to be *orthonormal* in (a, b).

Example: The functions $\left\{ \dfrac{\sin mx}{\sqrt{\pi}} \right\}$, $n = 1, 2, \ldots$, are orthonormal in $(-\pi, \pi)$. See Problem 8.23.

ORTHONORMAL SERIES

A series $\qquad c_1 \phi_1(x) + c_2 \phi_2(x) + \cdots \;=\; \displaystyle\sum_{k=1}^{\infty} c_k \phi_k(x)$ $\tag{13}$

where $c_1, c_2, \ldots$ are constants and $\phi_1(x), \phi_2(x), \ldots$ are orthonormal functions in (a, b) is called an *orthonormal series*. If

$$c_n \;=\; \int_a^b f(x)\, \phi_n(x)\, dx \tag{14}$$

then the series (*13*) is called the orthonormal series corresponding to $f(x)$. Because of the analogy with Fourier series, (*13*) is sometimes called a *generalized Fourier series* and (*14*) are the *generalized Fourier coefficients*.

If $f(x)$ and $\phi_n(x)$ belong to L^2, the coefficients (*14*) exist [see Problem 8.69].

PARSEVAL'S IDENTITY

Let

$$S_n(x) = \sum_{k=1}^{n} c_k \phi_k(x) \tag{15}$$

be the partial sums of the series (*13*) with coefficients (*14*). Then by Problem 8.27,

$$\|S_n(x) - f(x)\|^2 = \int_a^b |S_n(x) - f(x)|^2 \, dx = \int_a^b |f(x)|^2 \, dx - \sum_{k=1}^{n} c_k^2$$

Now if it is true that

$$\lim_{n \to \infty} \|S_n(x) - f(x)\| = 0$$

i.e. if $S_n(x)$ converges in the mean to $f(x)$, it follows that

$$\int_a^b |f(x)|^2 \, dx = \sum_{k=1}^{\infty} c_k^2 \tag{16}$$

which is called *Parseval's identity*. Conversely if Parseval's identity holds, then $S_n(x)$ converges in the mean to $f(x)$. The results can be summarized in the following

Theorem 8-7. The generalized Fourier series corresponding to $f(x)$ converges in the mean to $f(x)$ if and only if Parseval's identity is satisfied.

BESSEL'S INEQUALITY

Regardless of whether $\lim\limits_{n \to \infty} \|S_n - f\|$ is or is not zero, it will certainly be true that

$$\sum_{k=1}^{\infty} c_k^2 \leqq \int_a^b |f(x)|^2 \, dx \tag{17}$$

See Problem 8.28. This inequality is called *Bessel's inequality*. The case of equality corresponds to Parseval's identity.

APPROXIMATIONS IN THE LEAST SQUARE SENSE

If $f(x) \in L^2$, then we can think of the quantity

$$\frac{1}{b-a} \int_a^b \left| f(x) - \sum_{k=1}^{n} \alpha_k \phi_k(x) \right|^2 dx \tag{18}$$

as the *mean square error* of $f(x)$ from an approximating sum $\sum\limits_{k=1}^{\infty} \alpha_k \phi_k(x)$. The following theorem is of interest.

Theorem 8-8. The mean square error (*18*) is a minimum when the constants α_k are the generalized Fourier coefficients, i.e. when

$$\alpha_k = c_k = \int_a^b f(x) \phi_k(x) \, dx \tag{19}$$

Because of this theorem we say that $\sum\limits_{k=1}^{n} c_k \phi_k(x)$ approximates $f(x)$ in the *least square sense*. Note that Parseval's identity is satisfied if and only if the mean square error approaches zero as $n \to \infty$. See Problems 8.30 and 8.76.

COMPLETENESS OF ORTHONORMAL SETS

If in approximating a function $f(x) \in L^2$ by an orthonormal series $\sum_{k=1}^{\infty} c_k \phi_k(x)$ we fail to include one or more of the functions $\phi_k(x)$, Parseval's identity will not be satisfied. Because of this we adopt the following

Definition. An orthonormal set $\{\phi_k(x)\}$ is said to be *complete* if for all functions $f(x) \in L^2$ Parseval's identity is satisfied.

We have the following

Theorem 8-9. The set $\{\phi_k(x)\}$ is complete if and only if there is no function other than zero which is orthogonal to all the functions $\phi_k(x)$.

RIESZ-FISCHER THEOREM FOR GENERALIZED FOURIER SERIES

Theorem 8-10 [**Riesz-Fischer**]. Given an orthonormal set $\{\phi_k(x)\}$ in (a, b) and a set of constants c_k such that $\sum_{k=1}^{\infty} c_k^2$ converges, there exists a function $f(x) \in L^2$ such that the c_k are Fourier coefficients corresponding to $f(x)$, Parseval's identity is satisfied and the series $\sum_{k=1}^{\infty} c_k \phi_k(x)$ converges in the mean to $f(x)$.

For Fourier series this theorem takes the following form

Theorem 8-11. Let a_k, b_k be such that the series

$$\frac{a_0^2}{2} + \sum_{k=1}^{\infty} (a_k^2 + b_k^2) \tag{20}$$

converges. Then the trigonometric series

$$\frac{a_0}{2} + \sum_{k=1}^{\infty} (a_k \cos kx + b_k \sin kx) \tag{21}$$

is the Fourier series of some function $f(x) \in L^2$. Furthermore the partial sums of the series (21) converge in the mean to $f(x)$ and Parseval's identity (16) is satisfied.

Solved Problems

DEFINITION OF FOURIER SERIES

8.1. Show that if m and n are positive integers,

$$(a) \quad \int_{-\pi}^{\pi} \cos mx \cos nx \, dx = \begin{cases} 0, & m \neq n \\ \pi, & m = n \end{cases}$$

$$(b) \quad \int_{-\pi}^{\pi} \sin mx \sin nx \, dx = \begin{cases} 0, & m \neq n \\ \pi, & m = n \end{cases}$$

$$(c) \quad \int_{-\pi}^{\pi} \sin mx \cos nx \, dx = 0$$

(a) If $m \neq n$,

$$\int_{-\pi}^{\pi} \cos mx \cos nx \, dx = \frac{1}{2} \int_{-\pi}^{\pi} [\cos (m-n)x + \cos (m+n)x] \, dx$$

$$= \frac{1}{2} \left[\frac{\sin (m-n)x}{m-n} + \frac{\sin (m+n)x}{m+n} \right] \Big|_{-\pi}^{\pi} = 0$$

If $m = n$,

$$\int_{-\pi}^{\pi} \cos mx \cos nx \, dx \;=\; \int_{-\pi}^{\pi} \cos^2 mx \, dx$$

$$=\; \frac{1}{2} \int_{-\pi}^{\pi} (1 + \cos 2mx) \, dx$$

$$=\; \frac{1}{2}\left(x + \frac{\sin 2mx}{2m}\right)\Big|_{-\pi}^{\pi} \;=\; \pi$$

(b) If $m \neq n$,

$$\int_{-\pi}^{\pi} \sin mx \sin nx \, dx \;=\; \frac{1}{2} \int_{-\pi}^{\pi} [\cos(m-n)x - \cos(m+n)x] \, dx$$

$$=\; \frac{1}{2}\left[\frac{\sin(m-n)x}{m-n} + \frac{\sin(m+n)x}{m+n}\right]\Big|_{-\pi}^{\pi} \;=\; 0$$

If $m = n$,

$$\int_{-\pi}^{\pi} \sin mx \sin nx \, dx \;=\; \int_{-\pi}^{\pi} \sin^2 mx \, dx$$

$$=\; \frac{1}{2} \int_{-\pi}^{\pi} (1 - \cos 2mx) \, dx$$

$$=\; \frac{1}{2}\left(x - \frac{\sin 2mx}{2m}\right)\Big|_{-\pi}^{\pi} \;=\; \pi$$

(c) If $m \neq n$,

$$\int_{-\pi}^{\pi} \sin mx \cos nx \, dx \;=\; \frac{1}{2} \int_{-\pi}^{\pi} [\sin(m+n)x + \sin(m-n)x] \, dx$$

$$=\; \frac{1}{2}\left[\frac{-\cos(m+n)x}{m+n} - \frac{\cos(m-n)x}{m-n}\right]\Big|_{-\pi}^{\pi} \;=\; 0$$

If $m = n$,

$$\int_{-\pi}^{\pi} \sin mx \cos nx \, dx \;=\; \int_{-\pi}^{\pi} \sin mx \cos mx \, dx$$

$$=\; \frac{\sin^2 mx}{2m}\Big|_{-\pi}^{\pi} \;=\; 0$$

8.2. If the series $A + \sum_{n=1}^{\infty} (a_n \cos nx + b_n \sin nx)$ converges uniformly to $f(x)$ in $(-\pi, \pi)$, show that for $n = 1, 2, 3, \ldots$, (a) $a_n = \dfrac{1}{\pi} \displaystyle\int_{-\pi}^{\pi} f(x) \cos nx \, dx$, (b) $b_n = \dfrac{1}{\pi} \displaystyle\int_{-\pi}^{\pi} f(x) \sin nx \, dx$, (c) $A = a_0/2$.

(a) Multiplying

$$f(x) \;=\; A + \sum_{n=1}^{\infty} (a_n \cos nx + b_n \sin nx) \tag{1}$$

by $\cos mx$ and integrating from $-\pi$ to π, using Problem 8.1, we have

$$\int_{-\pi}^{\pi} f(x) \cos mx \, dx \;=\; A \int_{-\pi}^{\pi} \cos mx \, dx$$

$$+ \sum_{n=1}^{\infty} \left\{ a_n \int_{-\pi}^{\pi} \cos mx \cos nx \, dx + b_n \int_{-\pi}^{\pi} \cos mx \sin nx \, dx \right\}$$

$$=\; a_m \pi \quad \text{if } m = 1, 2, 3, \ldots$$

Thus

$$a_m \;=\; \frac{1}{\pi} \int_{-\pi}^{\pi} f(x) \cos mx \, dx \qquad m = 1, 2, 3, \ldots \tag{2}$$

(b) Multiplying (1) by $\sin mx$ and integrating from $-\pi$ to π, using Problem 8.1, we have

$$\int_{-\pi}^{\pi} f(x) \sin mx \, dx \;=\; A \int_{-\pi}^{\pi} \sin mx \, dx$$
$$+ \sum_{n=1}^{\infty} \left\{ a_n \int_{-\pi}^{\pi} \sin mx \cos nx \, dx \;+\; b_n \int_{-\pi}^{\pi} \sin mx \sin nx \, dx \right\}$$
$$=\; b_m \pi$$

Thus
$$b_m \;=\; \frac{1}{\pi} \int_{-\pi}^{\pi} f(x) \sin mx \, dx \qquad m = 1, 2, 3, \ldots$$

(c) Integration of (1) from $-\pi$ to π, yields

$$\int_{-\pi}^{\pi} f(x) \, dx \;=\; A \int_{-\pi}^{\pi} dx + \sum_{n=1}^{\infty} \left\{ a_n \int_{-\pi}^{\pi} \cos nx \, dx + b_n \int_{-\pi}^{\pi} \sin nx \, dx \right\} \;=\; 2\pi A$$

Thus
$$A \;=\; \frac{1}{2\pi} \int_{-\pi}^{\pi} f(x) \, dx \;=\; \frac{a_0}{2}$$

as seen by formally putting $m = 0$ in equation (2) of part (a).

Note that in all parts above, interchange of summation and integration is valid because the series is *assumed* to converge uniformly to $f(x)$ in $(-\pi, \pi)$. Even when this assumption is not warranted, the coefficients a_n and b_n as obtained above are called the *Fourier coefficients* corresponding to $f(x)$, and the corresponding trigonometric series with these values of a_n and b_n is then called the *Fourier series* corresponding to $f(x)$.

An important problem is to investigate conditions under which the series converges and if so whether it converges to $f(x)$.

8.3. Find the Fourier series corresponding to the function $f(x) = x^2$, $0 < x < 2\pi$, where $f(x)$ has period 2π outside of the interval $(0, 2\pi)$.

Although it is of course not necessary to graph the function in order to determine the Fourier series, it is sometimes useful to do so. The graph of $f(x)$ showing the periodicity is given in Fig. 8-1. Since $f(x)$ has period 2π, we can use $(0, 2\pi)$ in Problem 8.2 instead of $(-\pi, \pi)$.

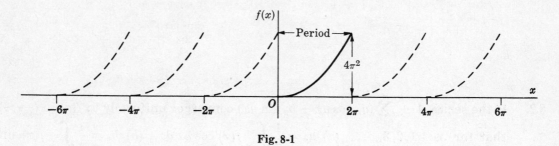

Fig. 8-1

The Fourier coefficients corresponding to $f(x)$ are given by the following.

$$a_n \;=\; \frac{1}{\pi} \int_{0}^{2\pi} f(x) \cos nx \, dx \;=\; \frac{1}{\pi} \int_{0}^{2\pi} x^2 \cos nx \, dx$$
$$=\; \frac{1}{\pi} \left\{ (x^2) \left(\frac{\sin nx}{n} \right) - (2x) \left(-\frac{\cos nx}{n^2} \right) + 2 \left(-\frac{\sin nx}{n^3} \right) \right\} \Big|_{0}^{2\pi} \;=\; \frac{4}{n^2}, \quad n = 1, 2, 3, \ldots$$

If $n = 0$,
$$a_0 \;=\; \frac{1}{\pi} \int_{0}^{2\pi} x^2 \, dx \;=\; \frac{8\pi^2}{3}$$

$$b_n \;=\; \frac{1}{\pi} \int_{0}^{2\pi} f(x) \sin nx \, dx \;=\; \frac{1}{\pi} \int_{0}^{2\pi} x^2 \sin nx \, dx$$
$$=\; \frac{1}{\pi} \left\{ (x^2) \left(-\frac{\cos nx}{n} \right) - (2x) \left(-\frac{\sin nx}{n^2} \right) + (2) \left(\frac{\cos nx}{n^3} \right) \right\} \Big|_{0}^{2\pi} \;=\; -\frac{4\pi}{n}$$

Then the required Fourier series is

$$\frac{a_0}{2} + \sum_{n=1}^{\infty} a_n \cos nx + b_n \sin nx$$

i.e.

$$\frac{4\pi^2}{3} + \sum_{n=1}^{\infty} \left(\frac{4}{n^2} \cos nx - \frac{4\pi}{n} \sin nx \right)$$

RIEMANN-LEBESGUE THEOREM

8.4. Prove the Riemann-Lebesgue theorem: If $f(x)$ is integrable in $(-\pi, \pi)$, then

$$(a)\ \lim_{n \to \infty} \int_{-\pi}^{\pi} f(x) \sin nx\, dx\ =\ 0, \qquad (b)\ \lim_{n \to \infty} \int_{-\pi}^{\pi} f(x) \cos nx\, dx\ =\ 0$$

(a) Let $x = u + \pi/n$ so that

$$\int_{-\pi}^{\pi} f(x) \sin nx\, dx\ =\ -\int_{-\pi+\pi/n}^{\pi+\pi/n} f\left(u + \frac{\pi}{n} \right) \sin nu\, du\ =\ -\int_{-\pi}^{\pi} f\left(u + \frac{\pi}{n} \right) \sin nu\, du$$

since we assume $f(x)$ has period 2π.

Thus

$$\int_{-\pi}^{\pi} f(x) \sin nx\, dx\ =\ -\int_{-\pi}^{\pi} f\left(x + \frac{\pi}{n} \right) \sin nx\, dx$$

and

$$2 \int_{-\pi}^{\pi} f(x) \sin nx\, dx\ =\ \int_{-\pi}^{\pi} \left[f(x) - f\left(x + \frac{\pi}{n} \right) \right] \sin nx\, dx$$

From this we find

$$\left| \int_{-\pi}^{\pi} f(x) \sin nx\, dx \right|\ =\ \frac{1}{2} \left| \int_{-\pi}^{\pi} \left[f(x) - f\left(x + \frac{\pi}{n} \right) \right] \sin nx\, dx \right|$$

$$\leq\ \frac{1}{2} \int_{-\pi}^{\pi} \left| f\left(x + \frac{\pi}{n} \right) - f(x) \right| dx$$

But by Problem 5.31, page 90, we have, since $f(x)$ is integrable,

$$\lim_{n \to \infty} \int_{-\pi}^{\pi} \left| f\left(x + \frac{\pi}{n} \right) - f(x) \right| dx\ =\ 0$$

Thus

$$\lim_{n \to \infty} \int_{-\pi}^{\pi} f(x) \sin nx\, dx\ =\ 0$$

(b) This can be proved in a manner similar to the method of part (a). See Problem 8.43.

CONVERGENCE OF FOURIER SERIES

8.5. Prove that $\dfrac{1}{2} + \displaystyle\sum_{n=1}^{M} \cos n\theta\ =\ \dfrac{\sin(M + \frac{1}{2})\theta}{2 \sin \frac{1}{2}\theta}$.

We have $\qquad \cos n\theta \sin \frac{1}{2}\theta\ =\ \frac{1}{2}[\sin(n + \frac{1}{2})\theta - \sin(n - \frac{1}{2})\theta]$

Then summing from $n = 1$ to M,

$$\sum_{n=1}^{M} \cos n\theta \sin \tfrac{1}{2}\theta\ =\ \sin \tfrac{1}{2}\theta \sum_{n=1}^{M} \cos n\theta$$

$$=\ \tfrac{1}{2}(\sin \tfrac{3}{2}\theta - \sin \tfrac{1}{2}\theta) + \tfrac{1}{2}(\sin \tfrac{5}{2}\theta - \sin \tfrac{3}{2}\theta) + \cdots + \tfrac{1}{2}(\sin(M + \tfrac{1}{2})\theta - \sin(M - \tfrac{1}{2})\theta)$$

$$=\ \tfrac{1}{2} \sin(M + \tfrac{1}{2})\theta - \tfrac{1}{2} \sin \tfrac{1}{2}\theta$$

Thus

$$\frac{1}{2} + \sum_{n=1}^{M} \cos n\theta\ =\ \frac{\sin(M + \frac{1}{2})\theta}{2 \sin \frac{1}{2}\theta}$$

8.6. Prove that $\quad$ (a) $\dfrac{1}{\pi} \displaystyle\int_0^\pi \dfrac{\sin{(M + \frac{1}{2})\theta}}{2 \sin{\frac{1}{2}\theta}}\, d\theta \;=\; \dfrac{1}{2},\quad$ (b) $\dfrac{1}{\pi} \displaystyle\int_{-\pi}^0 \dfrac{\sin{(M + \frac{1}{2})\theta}}{2 \sin{\frac{1}{2}\theta}}\, d\theta \;=\; \dfrac{1}{2}.$

(a) From Problem 8.5 we have on integrating from 0 to π,

$$\frac{1}{\pi} \int_0^\pi \frac{\sin{(M + \frac{1}{2})\theta}}{2 \sin{\frac{1}{2}\theta}}\, du \;=\; \frac{1}{\pi} \int_0^\pi \left[\frac{1}{2} + \sum_{n=1}^M \cos{n\theta} \right] d\theta$$

$$=\; \frac{1}{\pi}\left[\frac{\theta}{2} + \sum_{n=1}^M \frac{\sin{n\theta}}{n} \right]\Bigg|_0^\pi$$

$$=\; \frac{1}{2}$$

(b) From Problem 8.5 we have on integrating from $-\pi$ to 0,

$$\frac{1}{\pi} \int_{-\pi}^0 \frac{\sin{(M + \frac{1}{2})\theta}}{2 \sin{\frac{1}{2}\theta}}\, d\theta \;=\; \frac{1}{\pi} \int_{-\pi}^0 \left[\frac{1}{2} + \sum_{n=1}^M \cos{n\theta} \right] d\theta$$

$$=\; \frac{1}{\pi}\left[\frac{\theta}{2} + \sum_{n=1}^M \frac{\sin{n\theta}}{n} \right]\Bigg|_{-\pi}^0$$

$$=\; \frac{1}{2}$$

8.7. Show that the partial sums of a Fourier series are given by

$$S_M(x) \;=\; \frac{a_0}{2} + \sum_{n=1}^M (a_n \cos{nx} + b_n \sin{nx}) \;=\; \frac{1}{\pi} \int_{-\pi}^\pi f(x+u)\, \frac{\sin{(M + \frac{1}{2})u}}{2 \sin{\frac{1}{2}u}}\, du$$

Using the formulas for the Fourier coefficients, we have

$$a_n \cos{nx} + b_n \sin{nx} \;=\; \left(\frac{1}{\pi} \int_{-\pi}^\pi f(t) \cos{nt}\, dt \right) \cos{nx} + \left(\frac{1}{\pi} \int_{-\pi}^\pi f(t) \sin{nt}\, dt \right) \sin{nx}$$

$$=\; \frac{1}{\pi} \int_{-\pi}^\pi f(t)(\cos{nt} \cos{nx} + \sin{nt} \sin{nx})\, dt$$

$$=\; \frac{1}{\pi} \int_{-\pi}^\pi f(t) \cos{n(t-x)}\, dt$$

Also, $$\frac{a_0}{2} \;=\; \frac{1}{2\pi} \int_{-\pi}^\pi f(t)\, dt$$

Then $$S_M(x) \;=\; \frac{a_0}{2} + \sum_{n=1}^M (a_n \cos{nx} + b_n \sin{nx})$$

$$=\; \frac{1}{2\pi} \int_{-\pi}^\pi f(t)\, dt + \frac{1}{\pi} \sum_{n=1}^M \int_{-\pi}^\pi f(t) \cos{n(t-x)}\, dt$$

$$=\; \frac{1}{\pi} \int_{-\pi}^\pi f(t) \left[\frac{1}{2} + \sum_{n=1}^M \cos{n(t-x)} \right] dt$$

$$=\; \frac{1}{2\pi} \int_{-\pi}^\pi f(t)\, \frac{\sin{(M + \frac{1}{2})(t-x)}}{\sin{\frac{1}{2}(t-x)}}\, dt$$

using Problem 8.5. Letting $t - x = u$, we then find

$$S_M(x) \;=\; \frac{1}{2\pi} \int_{-\pi-x}^{\pi-x} f(x+u)\, \frac{\sin{(M + \frac{1}{2})u}}{\sin{\frac{1}{2}u}}\, du$$

Since the integrand has period 2π, we can replace the interval $(-\pi - x,\ \pi - x)$ by any other interval of length 2π, in particular $(-\pi, \pi)$. Thus we obtain the required result.

8.8. Prove that

$$S_M(x) = \frac{1}{2\pi} \int_0^\pi [f(x+t) + f(x-t)] \frac{\sin(M+\frac{1}{2})t}{\sin\frac{1}{2}t} \, dt$$

From Problem 8.7,

$$S_M(x) = \frac{1}{2\pi} \int_{-\pi}^\pi f(x+t) \frac{\sin(M+\frac{1}{2})t}{\sin\frac{1}{2}t} \, dt$$

$$= \frac{1}{2\pi} \left[\int_0^\pi f(x+t) \frac{\sin(M+\frac{1}{2})t}{\sin\frac{1}{2}t} \, dt + \int_{-\pi}^0 f(x+t) \frac{\sin(M+\frac{1}{2})t}{\sin\frac{1}{2}t} \, dt \right]$$

$$= \frac{1}{2\pi} \left[\int_0^\pi f(x+t) \frac{\sin(M+\frac{1}{2})t}{\sin\frac{1}{2}t} \, dt + \int_0^\pi f(x-u) \frac{\sin(M+\frac{1}{2})u}{\sin\frac{1}{2}u} \, du \right]$$

$$= \frac{1}{2\pi} \left[\int_0^\pi f(x+t) \frac{\sin(M+\frac{1}{2})t}{\sin\frac{1}{2}t} \, dt + \int_0^\pi f(x-t) \frac{\sin(M+\frac{1}{2})t}{\sin\frac{1}{2}t} \, dt \right]$$

$$= \frac{1}{2\pi} \int_0^\pi [f(x+t) + f(x-t)] \frac{\sin(M+\frac{1}{2})t}{\sin\frac{1}{2}t} \, dt$$

8.9. Prove that

$$S_M(x) - S(x) = \frac{1}{2\pi} \int_0^\pi [f(x+t) + f(x-t) - 2S(x)] \frac{\sin(M+\frac{1}{2})t}{\sin\frac{1}{2}t} \, dt$$

From Problem 8.6(a),

$$S(x) = \frac{1}{2\pi} \int_0^\pi 2S(x) \frac{\sin(M+\frac{1}{2})t}{\sin\frac{1}{2}t} \, dt$$

Then using Problem 8.8,

$$S_M(x) - S(x) = \frac{1}{2\pi} \int_0^\pi [f(x+t) + f(x-t)] \frac{\sin(M+\frac{1}{2})t}{\sin\frac{1}{2}t} \, dt - \frac{1}{2\pi} \int_0^\pi 2S(x) \frac{\sin(M+\frac{1}{2})t}{\sin\frac{1}{2}t} \, dt$$

$$= \frac{1}{2\pi} \int_0^\pi [f(x+t) + f(x-t) - 2S(x)] \frac{\sin(M+\frac{1}{2})t}{\sin\frac{1}{2}t} \, dt$$

8.10. Prove that convergence or divergence of a Fourier series corresponding to an integrable function $f(x)$ at any particular point x depends only on the values of $f(x)$ in a neighborhood of the point.

From Problem 8.8 we have for some fixed number δ such that $0 < \delta \leq \pi$,

$$S_M(x) = \frac{1}{2\pi} \int_0^\delta [f(x+t) + f(x-t)] \frac{\sin(M+\frac{1}{2})t}{\sin\frac{1}{2}t} \, dt$$

$$+ \frac{1}{2\pi} \int_\delta^\pi [f(x+t) + f(x-t)] \frac{\sin(M+\frac{1}{2})t}{\sin\frac{1}{2}t} \, dt$$

Now since $\dfrac{f(x+t) + f(x-t)}{\sin\frac{1}{2}t}$ is integrable in $\langle \delta, \pi \rangle$, it follows from the Riemann-Lebesgue theorem that the last integral approaches zero as $M \to \infty$. This shows that convergence or divergence of the Fourier series depends only on the integral over the interval $(0, \delta)$, i.e. on the values of $f(x)$ in a neighborhood of the point x.

8.11. Prove Theorem 8-1, page 131: The Fourier series corresponding to $f(x)$ converges to $S(x)$ if and only if

$$\lim_{M \to \infty} \int_0^\pi [f(x+t) + f(x-t) - 2S(x)] \frac{\sin(M+\frac{1}{2})t}{\sin\frac{1}{2}t} \, dt = 0$$

This follows at once from Problem 8.9 since in such case $\lim_{M \to \infty} S_M(x) = S(x)$.

8.12. Prove Theorem 8-2, page 131: The Fourier series corresponding to $f(x)$ converges to $S(x)$ if and only if for some fixed number δ such that $0 < \delta \leqq \pi$,

$$\lim_{M \to \infty} \int_0^\delta [f(x+t) + f(x-t) - 2S(x)] \frac{\sin(M+\frac{1}{2})t}{\sin\frac{1}{2}t}\, dt \;=\; 0$$

If we let $F(t) = f(x+t) + f(x-t) - 2S(x)$, then by Problem 8.11 the Fourier series converges to $S(x)$ if and only if

$$\lim_{M \to \infty} \int_0^\pi F(t)\, \frac{\sin(M+\frac{1}{2})t}{\sin\frac{1}{2}t}\, dt \;=\; 0$$

i.e. for some fixed number δ such that $0 < \delta \leqq \pi$,

$$\lim_{M \to \infty} \left[\int_0^\delta F(t)\, \frac{\sin(M+\frac{1}{2})t}{\sin\frac{1}{2}t}\, dt + \int_\delta^\pi F(t)\, \frac{\sin(M+\frac{1}{2})t}{\sin\frac{1}{2}t}\, dt \right] \;=\; 0$$

Now since the limit of the second integral is zero by the Riemann-Lebesgue theorem, the required result follows.

8.13. Prove Theorem 8-3, page 131: The Fourier series corresponding to $f(x)$ converges to $S(x)$ if and only if for some fixed number δ such that $0 < \delta \leqq \pi$,

$$\lim_{M \to \infty} \int_0^\delta [f(x+t) + f(x-t) - 2S(x)] \frac{\sin(M+\frac{1}{2})t}{t}\, dt \;=\; 0$$

Using Problem 8.12 we see that the Fourier series converges to $S(x)$ if and only if for some fixed number δ such that $0 < \delta \leqq \pi$,

$$\lim_{M \to \infty} \int_0^\delta F(t)\, \frac{\sin(M+\frac{1}{2})t}{\sin\frac{1}{2}t}\, dt \;=\; 0 \tag{1}$$

Now the function

$$\frac{1}{\sin\frac{1}{2}t} - \frac{1}{\frac{1}{2}t} \;=\; \frac{1}{\sin\frac{1}{2}t} - \frac{2}{t}$$

is integrable in the interval $(0, \delta)$ [see Problem 8.45]. Thus from the Riemann-Lebesgue theorem,

$$\lim_{M \to \infty} \int_0^\delta F(t)\left[\frac{1}{\sin\frac{1}{2}t} - \frac{2}{t} \right] \sin(M+\tfrac{1}{2})t\, dt \;=\; 0 \tag{2}$$

Then from (1) and (2) we see that

$$\lim_{M \to \infty} \int_0^\delta F(t)\, \frac{\sin(M+\frac{1}{2})t}{t}\, dt \;=\; 0$$

which proves the required result.

SUFFICIENT CONDITIONS FOR CONVERGENCE OF FOURIER SERIES

8.14. Prove Theorem 8-4, page 131 and thus establish *Dini's condition* for convergence of Fourier series: The Fourier series converges to $S(x)$ if for some fixed number δ such that $0 < \delta \leqq \pi$,

$$\int_0^\delta \frac{F(t)}{t}\, dt \text{ exists}$$

where $F(t) = f(x+t) + f(x-t) - 2S(x)$.

By the Riemann-Lebesgue theorem if $F(t)/t$ is integrable in $(0, \delta)$, i.e. if

$$\int_0^\delta \frac{F(t)}{t}\, dt \quad \text{exists}$$

then
$$\lim_{M \to \infty} \int_0^\delta F(t) \frac{\sin (M + \frac{1}{2})t}{t} dt = 0$$

But this is exactly the condition [Problem 8.13] that the Fourier series converges to $S(x)$ and so the required result follows.

8.15. Let $\psi(t)$ be a monotonic increasing function and suppose that $\lim_{t \to 0+} \psi(t) = 0$. Prove that if $0 < \delta \leqq \pi$,
$$\lim_{M \to \infty} \int_0^\delta \psi(t) \frac{\sin (M + \frac{1}{2})t}{t} dt = 0$$

By the second mean-value theorem [see Problems 6.78-80, page 114] we have for $0 < \delta_1 \leqq \pi$
$$\int_0^{\delta_1} \psi(t) \frac{\sin (M + \frac{1}{2})t}{t} dt = \psi(\delta_1) \int_\eta^{\delta_1} \frac{\sin (M + \frac{1}{2})t}{t} dt \qquad (1)$$

for some value η such that $0 < \eta < \delta_1$.

Letting $(M + \frac{1}{2})t = u$ in the integral on the right of (1),
$$\int_0^{\delta_1} \psi(t) \frac{\sin (M + \frac{1}{2})t}{t} dt = \psi(\delta_1) \int_{(M + \frac{1}{2})\eta}^{(M + \frac{1}{2})\delta_1} \frac{\sin u}{u} du \qquad (2)$$

Now the integral on the right of (2) is bounded, i.e. less than some positive number B, for all M, δ_1 and η [see Problem 8.47]. Then given $\epsilon > 0$, we can choose δ_1 small enough so that $\psi(\delta_1) < \epsilon/2B$. Thus
$$\left| \int_0^{\delta_1} \psi(t) \frac{\sin (M + \frac{1}{2})t}{t} dt \right| < \frac{\epsilon}{2} \qquad (3)$$

Also by the Riemann-Lebesgue theorem,
$$\lim_{M \to \infty} \int_{\delta_1}^\delta \psi(t) \frac{\sin (M + \frac{1}{2})t}{t} dt = 0$$

i.e. we can choose M large enough so that
$$\left| \int_{\delta_1}^\delta \psi(t) \frac{\sin (M + \frac{1}{2})t}{t} dt \right| < \frac{\epsilon}{2} \qquad (4)$$

Then from (3) and (4) it follows that if M is sufficiently large,
$$\left| \int_0^\delta \psi(t) \frac{\sin (M + \frac{1}{2})t}{t} dt \right| < \epsilon$$

which proves the required result.

8.16. Prove the result of Problem 8.15 if $\psi(t)$ is of bounded variation and $\lim_{t \to 0+} \psi(t) = 0$.

If $\psi(t)$ is of bounded variation, then it can be expressed as the difference $\psi_1(t) - \psi_2(t)$ of two monotonic increasing functions $\psi_1(t)$ and $\psi_2(t)$. Since $\lim_{t \to 0+} \psi(t) = 0$, we have $\lim_{t \to 0+} \psi_1(t) = \lim_{t \to 0+} \psi_2(t)$ and there is no loss of generality in assuming that this common limit is zero. Then by Problem 8.15,
$$\lim_{M \to \infty} \int_0^\delta \psi_1(t) \frac{\sin (M + \frac{1}{2})t}{t} dt = 0, \qquad \lim_{M \to \infty} \int_0^\delta \psi_2(t) \frac{\sin (M + \frac{1}{2})t}{t} dt = 0$$

and so by taking the difference of these two limits we obtain as required
$$\lim_{M \to \infty} \int_0^\delta \psi(t) \frac{\sin (M + \frac{1}{2})t}{t} dt = 0$$

8.17. Prove Theorem 8-5, page 131, and thus establish *Jordan's condition* for convergence of Fourier series: The Fourier series converges to $S(x) = \frac{1}{2}[f(x+0) + f(x-0)]$ if $f(t)$ is of bounded variation in a neighborhood of $t = x$.

If $f(t)$ is of bounded variation, so also is

$$F(t) = f(x+t) + f(x-t) - 2S(x) = f(x+t) + f(x-t) - f(x+0) - f(x-0)$$

Furthermore it is clear that $\lim_{t \to 0+} F(t) = 0$. Thus by Problem 8.16,

$$\lim_{M \to \infty} \int_0^{\delta} F(t) \frac{\sin (M + \frac{1}{2})t}{t} \, dt = 0$$

which shows that the Fourier series converges to $S(x) = \frac{1}{2}[f(x+0) + f(x-0)]$.

8.18. (a) Referring to Problem 8.3, page 136, prove that for $0 < x < 2\pi$,

$$x^2 = \frac{4\pi^2}{3} + \sum_{n=1}^{\infty} \left(\frac{4}{n^2} \cos nx - \frac{4\pi}{n} \sin nx \right)$$

(b) To what value does the series converge for $x = 0$ and $x = 2\pi$?

Since $f(x) = x^2$ is of bounded variation, the series converges to

$$\tfrac{1}{2}[f(x+0) + f(x-0)]$$

(a) Since any point of the interval $0 < x < 2\pi$ is a point of continuity, the series converges to $f(x) = x^2$. Thus for $0 < x < 2\pi$,

$$x^2 = \frac{4\pi^2}{3} + \sum_{n=1}^{\infty} \left(\frac{4}{n^2} \cos nx - \frac{4\pi}{n} \sin nx \right)$$

(b) The point $x = 0$ is a point of discontinuity and we have from the graph of Fig. 8-1 or from direct analytical considerations the fact that at $x = 0$,

$$\tfrac{1}{2}[f(x+0) + f(x-0)] = \tfrac{1}{2}(0 + 4\pi^2) = 2\pi^2$$

Thus at $x = 0$ the series converges to $2\pi^2$.

Similarly at the point $x = 2\pi$ the series converges to

$$\tfrac{1}{2}[f(x+0) + f(x-0)] = \tfrac{1}{2}(0 + 4\pi^2) = 2\pi^2$$

8.19. Prove that $\dfrac{1}{1^2} + \dfrac{1}{2^2} + \dfrac{1}{3^2} + \dfrac{1}{4^2} + \cdots = \dfrac{\pi^2}{6}$.

From the fact that the series of Problem 8.18 converges to $2\pi^2$ at $x = 0$, we have

$$2\pi^2 = \frac{4\pi^2}{3} + \sum_{n=1}^{\infty} \frac{4}{n^2} \qquad \text{or} \qquad \sum_{n=1}^{\infty} \frac{4}{n^2} = \frac{2\pi^2}{3}$$

Thus

$$\sum_{n=1}^{\infty} \frac{1}{n^2} = \frac{1}{1^2} + \frac{1}{2^2} + \frac{1}{3^2} + \frac{1}{4^2} + \cdots = \frac{\pi^2}{6}$$

INTEGRATION OF FOURIER SERIES

8.20. Prove Theorem 8-6, page 131.

We shall prove the result for $\alpha = 0$, $\beta = x$. Consider

$$g(x) = \int_0^x \left[f(u) - \frac{a_0}{2} \right] du \tag{1}$$

where a_k, b_k are the Fourier coefficients corresponding to $f(x)$, i.e.

$$a_k = \frac{1}{\pi} \int_0^{2\pi} f(x) \cos kx \, dx, \qquad b_k = \frac{1}{\pi} \int_0^{2\pi} f(x) \sin kx \, dx$$

Since $g(x)$ is continuous and of bounded variation [Theorem 6-21, page 98], it has a Fourier series which converges to $g(x)$, so that

$$g(x) = \frac{c_0}{2} + \sum_{k=1}^{\infty} (c_k \cos kx + d_k \sin kx) \tag{2}$$

where $\quad c_k = \frac{1}{\pi} \int_0^{2\pi} g(x) \cos kx \, dx, \qquad d_k = \frac{1}{\pi} \int_0^{2\pi} g(x) \sin kx \, dx$

Now using integration by parts with $g(x)$ given by (1), we have for $k = 1, 2, 3, \ldots$.

$$c_k = \frac{1}{\pi} \int_0^{2\pi} g(x) \cos kx \, dx = \frac{1}{\pi} \left[g(x) \frac{\sin kx}{k} \right]\Big|_0^{2\pi} - \frac{1}{k\pi} \int_0^{2\pi} g'(x) \sin kx \, dx$$

$$= -\frac{1}{k\pi} \int_0^{2\pi} \left[f(x) - \frac{a_0}{2} \right] \sin kx \, dx$$

$$= -\frac{1}{k\pi} \int_0^{2\pi} f(x) \sin kx \, dx = -\frac{b_k}{k}$$

$$d_k = \frac{1}{\pi} \int_0^{2\pi} g(x) \sin kx \, dx = \frac{1}{\pi} \left[-g(x) \frac{\cos kx}{k} \right]\Big|_0^{2\pi} + \frac{1}{k\pi} \int_0^{2\pi} g'(x) \cos kx \, dx$$

$$= \frac{1}{k\pi} \int_0^{2\pi} \left[f(x) - \frac{a_0}{2} \right] \cos kx \, dx$$

$$= \frac{1}{k\pi} \int_0^{2\pi} f(x) \cos kx \, dx = \frac{a_k}{k}$$

where we have used the fact that $g(0) = 0$, as is evident from (1), and $g(2\pi) = 0$ from the fact that $g(x)$ has period 2π.

Substituting the values of c_k and d_k in (2),

$$g(x) = \frac{c_0}{2} + \sum_{k=1}^{\infty} \frac{a_k \sin kx - b_k \cos kx}{k} \tag{3}$$

Letting $x = 0$, $\qquad 0 = \frac{c_0}{2} - \sum_{k=1}^{\infty} \frac{b_k}{k} \quad$ or $\quad \frac{c_0}{2} = \sum_{k=1}^{\infty} \frac{b_k}{k} \tag{4}$

so that $\qquad\qquad g(x) = \sum_{k=1}^{\infty} \frac{a_k \sin kx + b_k(1 - \cos kx)}{k} \tag{5}$

From (1) and (5),

$$\int_0^x f(u) \, du = \frac{a_0 x}{2} + \sum_{k=1}^{\infty} \frac{a_k \sin kx + b_k(1 - \cos kx)}{k}$$

But this is exactly what would be obtained if the Fourier series corresponding to $f(x)$, i.e.

$$\frac{a_0}{2} + \sum_{k=1}^{\infty} (a_k \cos kx + b_k \sin kx)$$

is integrated term by term from 0 to x, and so the required result is proved.

8.21. Show that $\dfrac{1}{1^4} + \dfrac{1}{2^4} + \dfrac{1}{3^4} + \dfrac{1}{4^4} + \cdots = \dfrac{\pi^4}{90}$.

By Theorem 8-6 we can integrate with respect to x the series of Problem 8.3 to obtain

$$\frac{x^3}{3} = \frac{4\pi^2 x}{3} + \sum_{n=1}^{\infty} \left(\frac{4}{n^3} \sin nx + \frac{4\pi}{n^2} \cos nx \right) + c_1 \tag{1}$$

where c_1 is the constant of integration. Then letting $x = 0$ we have, using Problem 8.19,

$$\sum_{n=1}^{\infty} \frac{4\pi}{n^2} + c_1 = 0 \quad \text{or} \quad c_1 = -4\pi \sum_{n=1}^{\infty} \frac{1}{n^2} = -\frac{2\pi^3}{3}$$

Thus (1) becomes

$$\frac{x^3}{3} - \frac{4\pi^2 x}{3} + \frac{2\pi^3}{3} \;=\; \sum_{n=1}^{\infty} \left(\frac{4}{n^3} \sin nx + \frac{4\pi}{n^2} \cos nx \right) \qquad (2)$$

Integrating both sides of (2) with respect to x,

$$\frac{x^4}{12} - \frac{2\pi^2 x^2}{3} + \frac{2\pi^3 x}{3} \;=\; \sum_{n=1}^{\infty} \left(\frac{-4 \cos nx}{n^4} + \frac{4\pi}{n^3} \sin nx \right) + c_2 \qquad (3)$$

Putting $x = 0$, $\qquad\qquad c_2 \;=\; \sum_{n=1}^{\infty} \frac{4}{n^4} \;=\; 4 \sum_{n=1}^{\infty} \frac{1}{n^4}$

which is four times the value of the series to be summed.

Another way must now be found to evaluate c_2. To do this we note that for $n = 1, 2, 3, \ldots$,

$$\int_0^{2\pi} \sin nx \, dx \;=\; 0, \qquad \int_0^{2\pi} \cos nx \, dx \;=\; 0$$

and are thus led to integrate both sides of (3) from 0 to 2π. We then obtain

$$\int_0^{2\pi} \left(\frac{x^4}{12} - \frac{2\pi^2 x^2}{3} + \frac{2\pi^3 x}{3} \right) dx \;=\; 0 + \int_0^{2\pi} c_2 \, dx$$

or, on carrying out the integration, $c_2 = 4\pi^4/90$, so that as required,

$$\sum_{n=1}^{\infty} \frac{1}{n^4} \;=\; \frac{\pi^4}{90}$$

This last method can also be used to determine c_1 in (1) without using the result of Problem 8.19.

8.22. Prove that the series

$$\frac{\sin x}{\ln 2} + \frac{\sin 2x}{\ln 3} + \frac{\sin 3x}{\ln 4} + \cdots$$

cannot be a Fourier series.

From Problem 8.20 we see that if the given series is a Fourier series, then $\displaystyle\sum_{k=1}^{\infty} \frac{b_k}{k}$ must converge. In the given series, however, $b_k = 1/\ln(k+1)$ and thus

$$\sum_{k=1}^{\infty} \frac{b_k}{k} \;=\; \sum_{k=1}^{\infty} \frac{1}{k \ln (k+1)} \;=\; \frac{1}{\ln 2} + \frac{1}{2 \ln 3} + \frac{1}{3 \ln 4} + \cdots$$

which diverges. Thus the given series cannot be a Fourier series.

ORTHONORMAL FUNCTIONS AND SERIES

8.23. Prove that the functions $\phi_n(x) = \dfrac{\sin nx}{\sqrt{\pi}}$, $n = 1, 2, 3, \ldots$ are orthonormal in $(-\pi, \pi)$.

The required result follows at once since

$$\int_{-\pi}^{\pi} \phi_m(x) \, \phi_n(x) \, dx \;=\; \frac{1}{\pi} \int_{-\pi}^{\pi} \sin mx \sin nx \, dx \;=\; \begin{cases} 0 & \text{if } m \neq n \\ 1 & \text{if } m = n \end{cases}$$

by Problem 8.1(b).

8.24. Let $\{\phi_n(x)\}$, $n = 1, 2, 3, \ldots$, be a set of orthonormal functions in (a, b). If $\displaystyle\sum_{n=1}^{\infty} c_n \phi_n(x)$ converges uniformly to $f(x)$, prove that

$$c_n \;=\; \int_a^b f(x) \, \phi_n(x) \, dx$$

If

$$f(x) = \sum_{n=1}^{\infty} c_n \, \phi_n(x)$$

then multiplying by $\phi_m(x)$,

$$f(x) \, \phi_m(x) = \sum_{n=1}^{\infty} c_n \, \phi_m(x) \, \phi_n(x)$$

Integrating from a to b, we have

$$\int_a^b f(x) \, \phi_m(x) \, dx = \sum_{n=1}^{\infty} c_n \int_a^b \phi_m(x) \, \phi_n(x) \, dx \tag{1}$$

where the integration of the series can be performed term by term since the series is supposed to be uniformly convergent in (a, b).

Since the set of functions $\{\phi_n(x)\}$ is orthonormal in (a, b), we have

$$\int_a^b \phi_m(x) \, \phi_n(x) \, dx = \begin{cases} 0 & \text{if } m \neq n \\ 1 & \text{if } m = n \end{cases}$$

so that (1) becomes as required,

$$c_m = \int_a^b f(x) \, \phi_m(x) \, dx$$

8.25. The *Hermite polynomials* $H_n(x)$, $n = 0, 1, 2, 3, \ldots$, are defined as polynomial solutions of degree n of the differential equation

$$H_n'' - 2xH_n' + 2nH_n = 0$$

Prove that the set of functions $\{e^{-x^2/2} H_n(x)\}$ is orthogonal in $(-\infty, \infty)$.

The differential equations satisfied by $H_m(x)$ and $H_n(x)$ are respectively

$$H_m'' - 2xH_m' + 2mH_m = 0 \tag{1}$$

$$H_n'' - 2xH_n' + 2nH_n = 0 \tag{2}$$

Multiplying the first equation by H_n, the second by H_m and subtracting, we find

$$H_n H_m'' - H_m H_n'' - 2x(H_n H_m' - H_m H_n') + (2m - 2n)H_m H_n = 0$$

so that

$$(2m - 2n)H_m H_n = H_m H_n'' - H_n H_m'' - 2x(H_m H_n' - H_n H_m')$$

or

$$(2m - 2n)H_m H_n = \frac{d}{dx}(H_m H_n' - H_n H_m') - 2x(H_m H_n' - H_n H_m')$$

Multiplying by e^{-x^2}, we can write this as

$$(2m - 2n)e^{-x^2} H_m H_n = \frac{d}{dx}[e^{-x^2}(H_m H_n' - H_n H_m')]$$

Then integrating from $-\infty$ to ∞,

$$(2m - 2n)\int_{-\infty}^{\infty} e^{-x^2} H_m H_n \, dx = e^{-x^2}(H_m H_n' - H_n H_m')\Big|_{-\infty}^{\infty} = 0$$

Thus if $m \neq n$,

$$\int_{-\infty}^{\infty} e^{-x^2} H_m H_n \, dx = 0 \tag{3}$$

which is the same as saying that the set $\{e^{-x^2/2} H_n(x)\}$ is orthogonal in $(-\infty, \infty)$.

8.26. If $\displaystyle\sum_{n=1}^{\infty} c_n H_n(x)$, where $H_n(x)$ are the Hermite polynomials of Problem 8.25, converges uniformly to $f(x)$, prove that

$$c_n = \frac{\displaystyle\int_{-\infty}^{\infty} e^{-x^2} f(x)\, H_n(x)}{\displaystyle\int_{-\infty}^{\infty} e^{-x^2} [H_n(x)]^2\, dx}$$

We have $\qquad\qquad f(x) = \displaystyle\sum_{n=1}^{\infty} c_n H_n(x)$

so that by equation (*3*) of Problem 8.25,

$$\int_{-\infty}^{\infty} e^{-x^2} f(x)\, H_m(x)\, dx = \sum_{n=1}^{\infty} c_n \int_{-\infty}^{\infty} e^{-x^2} H_m(x)\, H_n(x)\, dx$$

$$= c_m \int_{-\infty}^{\infty} e^{-x^2} [H_m(x)]^2\, dx \qquad (4)$$

where the term by term integration is justified by the uniform convergence of the series. Solving for c_m, we arrive at the required result.

FOURIER SERIES IN L^2 SPACES

8.27. Let $\{\phi_n(x)\}$ be an orthonormal set of functions $\in L^2\,[a, b]$ and let $S_n(x) = \displaystyle\sum_{k=1}^{n} c_k \phi_k(x)$ where $c_k = \displaystyle\int_a^b f(x)\,\phi_k(x)\, dx$. Prove that for any function $f(x) \in L^2\,[a, b]$,

$$\|S_n(x) - f(x)\| = \int_a^b |S_n(x) - f(x)|^2\, dx = \int_a^b |f(x)|^2\, dx - \sum_{k=1}^{n} c_k^2$$

We have,

$$\|S_n(x) - f(x)\| = \int_a^b |f(x) - S_n(x)|^2\, dx = \int_a^b \left| f(x) - \sum_{k=1}^{n} c_k \phi_k(x) \right|^2 dx$$

$$= \int_a^b \left\{ |f(x)|^2 - 2 \sum_{k=1}^{n} c_k\, f(x)\, \phi_k(x) + \sum_{k=1}^{n} \sum_{j=1}^{n} c_j c_k\, \phi_j(x)\, \phi_k(x) \right\} dx$$

$$= \int_a^b |f(x)|^2\, dx - 2 \sum_{k=1}^{n} c_k \int_a^b f(x)\, \phi_k(x)\, dx + \sum_{k=1}^{n} \sum_{j=1}^{n} c_j c_k \int_a^b \phi_j(x)\, \phi_k(x)\, dx$$

$$= \int_a^b |f(x)|^2\, dx - 2 \sum_{k=1}^{n} c_k^2 + \sum_{k=1}^{n} c_k^2$$

$$= \int_a^b |f(x)|^2\, dx - \sum_{k=1}^{n} c_k^2$$

where we have used the results

$$c_k = \int_a^b f(x)\,\phi_k(x)$$

$$\int_a^b \phi_j(x)\,\phi_k(x)\, dx = \begin{cases} 0, & j \ne k \\ 1, & j = k \end{cases}$$

8.28. From Problem 8.27, prove that

$$\sum_{k=1}^{\infty} c_k^2 \leqq \int_a^b |f(x)|^2\, dx$$

We have from the result of Problem 8.27,

$$\int_a^b |f(x) - S_n(x)|^2\, dx = \int_a^b |f(x)|^2\, dx - \sum_{k=1}^{n} c_k^2 \qquad (1)$$

Then since the left side is non-negative, it follows that

$$\sum_{k=1}^{n} c_k^2 \;\le\; \int_a^b |f(x)|^2\,dx \qquad\qquad (2)$$

Taking the limit as $n \to \infty$ and noting that the right side of (2) is independent of n, we see that

$$\sum_{k=1}^{\infty} c_k^2 \;\le\; \int_a^b |f(x)|^2\,dx$$

This inequality is called *Bessel's inequality*.

8.29. Prove that if

$$\lim_{n \to \infty} \|S_n(x) - f(x)\| \;=\; 0$$

where $S_n(x)$ and $f(x)$ are as defined in Problem 8.27, then

$$\sum_{k=1}^{\infty} c_k^2 \;=\; \int_a^b |f(x)|^2\,dx$$

This follows at once from Problem 8.27, since if

$$\lim_{n \to \infty} \|S_n(x) - f(x)\| \;=\; 0$$

then

$$\lim_{n \to \infty} \left[\int_a^b |f(x)|^2\,dx \;-\; \sum_{k=1}^{n} c_k^2 \right] \;=\; 0$$

i.e.

$$\int_a^b |f(x)|^2\,dx \;=\; \sum_{k=1}^{\infty} c_k^2$$

The result is called *Parseval's identity*. If Parseval's identity is satisfied for all functions $f(x) \in L^2$, we call the set of orthonormal functions $\{\phi_k(x)\}$ *complete*.

8.30. The *mean square error* of $f(x)$ from an approximating sum $\displaystyle\sum_{k=1}^{n} \alpha_k\,\phi_k(x)$ is given by

$$\frac{1}{b-a} \int_a^b \left| f(x) \;-\; \sum_{k=1}^{n} \alpha_k\,\phi_k(x) \right|^2 dx$$

Prove that (a) the mean square error is a minimum when the constants α_k are given by the generalized Fourier coefficients

$$\alpha_k \;=\; c_k \;=\; \int_a^b f(x)\,\phi_k(x)\,dx$$

and that (b) Parseval's identity is satisfied if and only if the mean square error approaches zero as $n \to \infty$.

(a) We have as in Problem 8.27,

$$\int_a^b \left| f(x) - \sum_{k=1}^{n} \alpha_k\,\phi_k(x) \right|^2 dx$$

$$= \int_a^b |f(x)|^2\,dx \;-\; 2\sum_{k=1}^{n} \alpha_k \int_a^b f(x)\,\phi_k(x)\,dx \;+\; \sum_{k=1}^{n}\sum_{j=1}^{n} \alpha_j \alpha_k \int_a^b \phi_j(x)\,\phi_k(x)\,dx$$

$$= \int_a^b |f(x)|^2\,dx \;-\; 2\sum_{k=1}^{n} \alpha_k c_k \;+\; \sum_{k=1}^{n} \alpha_k^2$$

$$= \int_a^b |f(x)|^2\,dx \;+\; \sum_{k=1}^{n} (\alpha_k^2 - 2\alpha_k c_k)$$

$$= \int_a^b |f(x)|^2\,dx \;-\; \sum_{k=1}^{n} c_k^2 \;+\; \sum_{k=1}^{n} (\alpha_k - c_k)^2$$

From this it is clear that a minimum is obtained when

$$\alpha_k = c_k = \int_a^b f(x)\,\phi_k(x)\,dx$$

For another method of obtaining the minimum, see Problem 8.76.

(b) This follows from part (a) and the result of Problem 8.29.

RIESZ-FISCHER THEOREM FOR GENERALIZED FOURIER SERIES

8.31. Let $\{\phi_k(x)\}$ be an orthonormal set in (a, b) and let c_k be a set of constants such that $\sum\limits_{k=1}^{\infty} c_k^2$ converges. Prove that there exists a function $f(x) \in L^2$ such that (a) the partial sums $\sum\limits_{k=1}^{n} c_k \phi_k(x)$ converge in the mean to $f(x)$ and that (b) the c_k are Fourier coefficients corresponding to $f(x)$.

(a) Let $S_n = \sum\limits_{k=1}^{n} c_k\,\phi_k(x)$ denote the nth partial sum of the series $\sum\limits_{k=1}^{\infty} c_k\phi_k(x)$. We have for $m > n$,

$$
\begin{aligned}
||S_m - S_n||^2 &= \int_a^b |S_m(x) - S_n(x)|^2\,dx \\
&= \int_a^b \left| \sum_{k=1}^{m} c_k\,\phi_k(x) - \sum_{k=1}^{n} c_k\,\phi_k(x) \right|^2 dx \\
&= \int_a^b \left| \sum_{n+1}^{m} c_k\,\phi_k(x) \right|^2 dx \\
&= \int_a^b \left\{ \sum_{k=n+1}^{m} \sum_{j=n+1}^{m} c_j c_k\,\phi_j(x)\,\phi_k(x) \right\} dx \\
&= \sum_{k=n+1}^{m} \sum_{j=n+1}^{m} c_j c_k \int_a^b \phi_j(x)\,\phi_k(x)\,dx \\
&= \sum_{k=n+1}^{m} c_k^2
\end{aligned}
$$

Now since $\sum\limits_{k=1}^{\infty} c_k^2$ converges, it follows that

$$\lim_{\substack{m \to \infty \\ n \to \infty}} \sum_{k=n+1}^{m} c_k^2 = 0$$

i.e.

$$\lim_{\substack{m \to \infty \\ n \to \infty}} ||S_m - S_n|| = 0$$

Thus by the Riesz-Fischer theorem there exists a function $f(x) \in L^2$ such that $S_n(x) = \sum\limits_{k=1}^{n} c_k\phi_k(x)$ converges in the mean to $f(x)$ as $n \to \infty$.

(b) By part (a), since $S_n(x)$ converges in the mean to $f(x)$ as $n \to \infty$, it follows from Problem 7.52, page 128, that

$$\lim_{n \to \infty} \int_a^b S_n(x)\,\phi_p(x)\,dx = \int_a^b f(x)\,\phi_p(x)\,dx \tag{1}$$

But

$$
\begin{aligned}
\int_a^b S_n(x)\,\phi_p(x)\,dx &= \int_a^b \left\{ \sum_{k=1}^{n} c_k\,\phi_k(x)\,\phi_p(x) \right\} dx \\
&= \sum_{k=1}^{n} c_k \int_a^b \phi_k(x)\,\phi_p(x)\,dx = c_p
\end{aligned}
$$

Thus (1) yields

$$c_p = \int_a^b f(x)\,\phi_p(x)\,dx \tag{2}$$

i.e. the c_k are Fourier coefficients corresponding to $f(x)$.

8.32. Prove that under the conditions specified in Problem 8.31 Parseval's identity holds, i.e.

$$\int_a^b |f(x)|^2\, dx \;=\; \sum_{k=1}^{\infty} c_k^2$$

By Problem 8.31 the partial sums $S_n(x)$ converge in the mean to $f(x)$ as $n \to \infty$. Thus by Problem 7.15, page 123,

$$\lim_{n\to\infty} \int_a^b |S_n(x)|^2\, dx \;=\; \int_a^b |f(x)|^2\, dx \tag{1}$$

But

$$\int_a^b |S_n(x)|^2\, dx \;=\; \int_a^b \left| \sum_{k=1}^n c_k\, \phi_k(x) \right|^2 dx$$

$$=\; \int_a^b \left\{ \sum_{k=1}^n \sum_{j=1}^n c_j c_k\, \phi_j(x)\, \phi_k(x) \right\} dx$$

$$=\; \sum_{k=1}^n \sum_{j=1}^n c_j c_k \int_a^b \phi_j(x)\, \phi_k(x)\, dx$$

$$=\; \sum_{k=1}^n c_k^2$$

Thus by (1),

$$\lim_{n\to\infty} \sum_{k=1}^n c_k^2 \;=\; \int_a^b |f(x)|^2\, dx$$

or

$$\sum_{k=1}^{\infty} c_k^2 \;=\; \int_a^b |f(x)|^2\, dx \tag{2}$$

Supplementary Problems

DEFINITION OF FOURIER SERIES

8.33. Let $f(x) = \begin{cases} x, & 0 < x < \pi \\ -x, & -\pi < x < 0 \end{cases}$ where $f(x)$ has period 2π. Obtain the Fourier series for $f(x)$ and draw the graph of $f(x)$. *Ans.* $2\left(\dfrac{\sin x}{1} - \dfrac{\sin 2x}{2} + \dfrac{\sin 3x}{3} - \dfrac{\sin 4x}{4} + \cdots \right)$

8.34. Show that the Fourier series for $f(x) = \cos px$, $-\pi \leqq x \leqq \pi$, where the period is 2π and where $p \neq 0, \pm 1, \pm 2, \ldots$ is given by

$$\frac{\sin p\pi}{\pi} \left(\frac{1}{p} - \frac{2p}{p^2 - 1^2} \cos x + \frac{2p}{p^2 - 2^2} \cos 2x - \frac{2p}{p^2 - 3^2} \cos 3x + \cdots \right)$$

8.35. Extend the results of (a) Problem 8.1 and (b) Problem 8.2 to the interval $(c, c + 2\pi)$ where c is any real number.

8.36. Let $f(x)$ be Lebesgue integrable in $(-l, l)$ or more generally in $(c, c + 2l)$ where $l > 0$ and c is any real number. Suppose that $f(x)$ is periodic with period $2l$ outside of this interval. Show that the Fourier series corresponding to $f(x)$ is

$$\frac{a_0}{2} + \sum_{n=1}^{\infty} \left(a_n \cos \frac{n\pi x}{l} + b_n \sin \frac{n\pi x}{l} \right)$$

where the Fourier coefficients are given by

$$a_n \;=\; \frac{1}{l} \int_c^{c+2l} f(x) \cos \frac{n\pi x}{l}\, dx, \qquad b_n \;=\; \frac{1}{l} \int_c^{c+2l} f(x) \sin \frac{n\pi x}{l}\, dx$$

8.37. Prove that if $f(x)$ is an *odd function* in $(-l, l)$ [i.e. $f(-x) = -f(x)$], then the Fourier series corresponding to $f(x)$ is

$$\sum_{n=1}^{\infty} b_n \sin \frac{n\pi x}{l} \quad \text{where} \quad b_n = \frac{2}{l} \int_0^l f(x) \sin \frac{n\pi x}{l} \, dx$$

The series is often called a *Fourier sine series*.

8.38. Prove that if $f(x)$ is an *even function* in $(-l, l)$ [i.e. $f(-x) = f(x)$], then the Fourier series corresponding to $f(x)$ is

$$\frac{a_0}{2} + \sum_{n=1}^{\infty} a_n \cos \frac{n\pi x}{l} \quad \text{where} \quad a_n = \frac{2}{l} \int_0^l f(x) \cos \frac{n\pi x}{l} \, dx$$

The series is often called a *Fourier cosine series*.

8.39. Show that the Fourier series corresponding to $f(x) = \begin{cases} 2x, & 0 \leq x < 3 \\ 0, & -3 < x < 0 \end{cases}$ where the period is 6 is given by

$$\frac{3}{2} + \sum_{n=1}^{\infty} \left[\frac{6(\cos n\pi - 1)}{n^2\pi^2} \cos \frac{n\pi x}{3} - \frac{6 \cos n\pi}{n\pi} \sin \frac{n\pi x}{3} \right]$$

8.40. Find a Fourier sine series corresponding to $f(x) = \cos x$, $0 < x < \pi$.   *Ans.* $\dfrac{8}{\pi} \sum_{n=1}^{\infty} \dfrac{n \sin 2nx}{4n^2 - 1}$

8.41. If the trigonometric series (*1*), page 130, converges almost everywhere to $f(x)$ in $(-\pi, \pi)$ and if the partial sums of the series are uniformly bounded, prove that (*a*) $f(x)$ is integrable in $(-\pi, \pi)$ and (*b*) a_n and b_n are given by the formulas (*2*), page 130. [*Hint*: Use Lebesgue's dominated convergence theorem].

THE RIEMANN-LEBESGUE THEOREM AND CONVERGENCE OF FOURIER SERIES

8.42. Prove that $\displaystyle\lim_{n \to \infty} \int_0^{\pi} \frac{\sin nx}{x^2 + 1} \, dx = 0$ in two different ways.

8.43. Prove that $\displaystyle\lim_{n \to \infty} \int_{-\pi}^{\pi} f(x) \cos nx \, dx = 0$ where $f(x)$ is integrable in $(-\pi, \pi)$ and thus complete the proof in Problem 8.4.

8.44. In the proof of the Riemann-Lebesgue theorem in Problem 8.4, does n have to be an integer? Explain.

8.45. Prove that the function $\dfrac{1}{\sin(t/2)} - \dfrac{2}{t}$ is integrable in $(0, \delta)$ and thus complete the proof in Problem 8.13.

8.46. Prove that if $f(x)$ is integrable in $(-\pi, \pi)$ and $f'(x)$ exists in $(-\pi, \pi)$, then the Fourier series corresponding to $f(x)$ converges to $f(x)$ at each point x in $(-\pi, \pi)$.

8.47. Prove that $\displaystyle\int_{(M+\frac{1}{2})\eta}^{(M+\frac{1}{2})\delta_1} \frac{\sin u}{u} \, du$ is bounded for all positive values of M, δ_1 and η and thus complete the proof in Problem 8.15.

8.48. Prove that a Fourier series corresponding to $f(x)$ will converge to $f(x)$ if $f(x)$ satisfies the condition $|f(x + u) - f(x)| < K |u|^p$ for $|u| < r$ where K, p and r are given positive constants. Discuss the relationship with Probem 8.46.

8.49. Prove that if $f(x)$ and $f'(x)$ are both piecewise continuous in $(-\pi, \pi)$, then the Fourier series for $f(x)$ converges to $\frac{1}{2}[f(x + 0) + f(x - 0)]$.

8.50. Prove that if $f(x)$ has only a finite number of discontinuities in $(-\pi, \pi)$ and if it has only a finite number of maxima and minima in $(-\pi, \pi)$, then the Fourier series for $f(x)$ converges to $\frac{1}{2}[f(x+0) + f(x-0)]$. The conditions indicated are often called *Dirichlet conditions*.

8.51. Is continuity of a function sufficient to assure the convergence of a Fourier series for $f(x)$ to $f(x)$? Can you justify your conclusion?

8.52. (a) To what value does the series of Problem 8.33 converge at $x = 0, \pi/2, \pi, -\pi$? (b) Does the series converge to $f(x)$? Explain. (c) Using Fourier series methods, prove that

$$1 - \frac{1}{3} + \frac{1}{5} - \frac{1}{7} + \cdots = \frac{\pi}{4}$$

8.53. Use Problem 8.34 to prove that if $p \neq 0, \pm 1, \pm 2, \ldots,$

$$\pi \cot p\pi = \frac{1}{p} + 2p \sum_{n=1}^{\infty} \frac{1}{p^2 - n^2}$$

8.54. Prove that for $0 \leqq x \leqq \pi$,

(a) $x(\pi - x) = \dfrac{\pi^2}{6} - \left(\dfrac{\cos 2x}{1^2} + \dfrac{\cos 4x}{2^2} + \dfrac{\cos 6x}{3^2} + \cdots \right)$

(b) $x(\pi - x) = \dfrac{8}{\pi} \left(\dfrac{\sin x}{1^3} + \dfrac{\sin 3x}{3^3} + \dfrac{\sin 5x}{5^3} + \cdots \right)$

8.55. Prove that if the function

$$G(t) = \frac{1}{t} \int_0^t [f(x+u) + f(x-u) - 2S(x)]\, du$$

is of bounded variation for $t > 0$, then the Fourier series corresponding to $f(x)$ is convergent and that it converges to $S(x)$ if $\lim_{t \to 0} G(t) = 0$. This test is called *de la Vallée Poussin's test*.

8.56. Show that the test of Problem 8.55 is satisfied if either Dini's or Jordan's test is satisfied.

INTEGRATION OF FOURIER SERIES

8.57. Prove that $\dfrac{1}{1^6} + \dfrac{1}{2^6} + \dfrac{1}{3^6} + \dfrac{1}{4^6} + \cdots = \dfrac{\pi^6}{945}$.

8.58. Prove that $\displaystyle\sum_{n=1}^{\infty} \frac{1}{n \ln (n+1)}$ diverges and thus complete the proof in Problem 8.22.

8.59. Prove that $\displaystyle\sum_{n=2}^{\infty} \frac{\sin nx}{(\ln n)^\alpha}$, $0 < \alpha \leqq 1$, cannot be a Fourier series.

8.60. Prove that if the Fourier series corresponding to $f(x)$ is

$$\frac{a_0}{2} + \sum_{n=1}^{\infty} (a_n \cos nx + b_n \sin nx)$$

and $\phi(x)$ is of bounded variation, then in any interval (α, β)

$$\int_\alpha^\beta f(x)\,\phi(x)\,dx = \frac{a_0}{2} \int_\alpha^\beta \phi(x)\,dx + \sum_{n=1}^{\infty} \left[a_n \int_\alpha^\beta \phi(x) \cos nx\,dx + b_n \int_\alpha^\beta \phi(x) \sin nx\,dx \right]$$

8.61. Discuss the result of Problem 8.60 if $\alpha = -\pi$, $\beta = \pi$ in terms of a Fourier series corresponding to $\phi(x)$.

8.62. Can the argument of Problem 8.22 be used to show that

$$\frac{\cos x}{\ln 2} + \frac{\cos 2x}{\ln 3} + \frac{\cos 3x}{\ln 4} + \cdots$$

is not a Fourier series? Explain.

ORTHONORMAL FUNCTIONS AND SERIES

8.63. Let $\phi_0(x) = 1/\sqrt{2\pi}$, $\phi_{2k-1}(x) = \dfrac{\sin kx}{\sqrt{\pi}}$, $\phi_{2k}(x) = \dfrac{\cos kx}{\sqrt{\pi}}$ where $k = 1, 2, 3, \ldots$. Prove that the set $\{\phi_n(x)\}$, $n = 0, 1, 2, 3, \ldots$, is orthonormal (a) in $(-\pi, \pi)$, (b) in $(c, c+2\pi)$ where c is any constant.

8.64. Use the method of Problem 8.25 to prove the orthogonality of the functions $\{\sin nx\}$, $n = 1, 2, \ldots$ in $(-\pi, \pi)$. [*Hint:* If $y_n(x) = \sin nx$, show that $y_n'' + n^2 y_n = 0$, $y_n(-\pi) = y_n(\pi) = 0$.]

8.65. The *Legendre polynomials* $P_n(x)$, $n = 0, 1, 2, 3, \ldots$, of degree n are solutions of the differential equation
$$(1 - x^2)y'' - 2xy' + n(n+1)y = 0$$
Prove that the functions are orthogonal in $(-1, 1)$.

8.66. The *Laguerre polynomials* $L_n(x)$, $n = 0, 1, 2, 3, \ldots$, of degree n are solutions of the differential equation
$$xy'' + (1 - x)y' + ny = 0$$
Prove that $\{e^{-x/2} L_n(x)\}$ is orthogonal in $(0, \infty)$.

8.67. Show how to expand a function $f(x)$ into a series of (a) Legendre polynomials and (b) Laguerre polynomials.

8.68. The function $\text{sgn}(x)$ [read "signum x"] is defined as 1 if $x > 0$, 0 if $x = 0$ and -1 if $x < 0$. Show that the functions $\{\text{sgn}(\sin 2^n \pi x)\}$, $n = 0, 1, 2, \ldots$, called *Rademacher functions*, are orthonormal in $(0, 1)$.

8.69. Prove that if $f(x)$ and $\phi_n(x)$ belong to L^2, then the coefficients (*14*), page 132, exist.

8.70. A set of complex functions $\{\phi_n(x)\}$ is called orthonormal in (a, b) if
$$\int_a^b \phi_m(x)\,\overline{\phi_n(x)}\,dx = \begin{cases} 0 & \text{if } m \neq n \\ 1 & \text{if } m = n \end{cases}$$
where the bar over a function denotes complex conjugate.

(a) Show that this definition reduces to the ordinary one in case the functions are real.

(b) Show that the set of functions $\left\{\dfrac{e^{inx}}{\sqrt{2\pi}}\right\}$ where n is any integer is orthonormal in $(-\pi, \pi)$.

(c) If $\displaystyle\sum_{k=-\infty}^{\infty} c_k e^{ikx}$ is the Fourier series for $f(x)$, prove that
$$c_k = \frac{1}{2\pi} \int_{-\pi}^{\pi} f(x)\, e^{-ikx}\, dx$$
This is called the *complex form of Fourier series*.

8.71. Consider the set of functions $1, x, x^2, x^3, \ldots$. From these form a new set of functions $\alpha_1, \alpha_2 + \alpha_3 x, \alpha_4 + \alpha_5 x + \alpha_6 x^2, \ldots$ where the nth function in the new set is obtained by multiplying each of the first n members of the first set by a constant and then adding. (a) Determine the constants $\alpha_1, \alpha_2, \ldots$ so that the functions will be orthonormal in $(-1, 1)$ and (b) show that these polynomials satisfy Legendre's differential equation [the actual Legendre polynomials have the property that $P_n(1) = 1$]. See Problem 8.65.

FOURIER SERIES IN L^2 SPACES AND THE RIESZ-FISCHER THEOREM

8.72. Let $f(x) \in L^2\,[-\pi, \pi]$ have the Fourier series
$$\frac{a_0}{2} + \sum_{n=1}^{\infty} (a_n \cos nx + b_n \sin nx)$$
Prove that Parseval's identity is
$$\frac{a_0^2}{2} + \sum_{n=1}^{\infty} (a_n^2 + b_n^2) = \int_{-\pi}^{\pi} |f(x)|^2\, dx$$
and thus prove that the series on the left converges.

8.73. If $f(x) \in L^2 \ [-\pi, \pi]$, use Problem 8.72 to prove that

$$\lim_{n \to \infty} \int_{-\pi}^{\pi} f(x) \cos nx \, dx \ = \ 0, \qquad \lim_{n \to \infty} \int_{-\pi}^{\pi} f(x) \sin nx \, dx \ = \ 0$$

Do these results prove the Riemann-Lebesgue theorem of page 130? Explain.

8.74. Verify Parseval's identity corresponding to the function of Problem 8.18.

8.75. Use Parseval's identity corresponding to the function of Problem 8.34 to find the sum of the series

$$\sum_{n=1}^{\infty} \frac{1}{(n^2 - p^2)^2}$$

8.76. Work Problem 8.30 by using partial differentiation to determine the minimum value of the least square error.

8.77. Determine which if any of the following series are Fourier series corresponding to a function in $L^2 \ [-\pi, \pi]$:

(a) $\displaystyle\sum_{n=1}^{\infty} \frac{\sin nx}{\sqrt{n}}$,   (b) $\displaystyle\sum_{n=1}^{\infty} \frac{\sin nx}{2n - 1}$,   (c) $\displaystyle\sum_{n=1}^{\infty} \frac{\cos nx}{\sqrt{n \ln(n+1)}}$.

8.78. Prove that if $f(x) \in L^2$ and $g(x) \in L^2$ have the same Fourier coefficients, then $f(x) = g(x)$ almost everywhere.

8.79. Is it possible for a given trigonometric series to be a Fourier series corresponding to (a) a function in L^2 but not in L, (b) a function in L but not in L^2? Explain.

8.80. Prove Theorem 8-9, page 134.

Appendix A

The Riemann Integral

DEFINITION OF THE RIEMANN INTEGRAL

Let $f(x)$ be defined and bounded in $[a, b]$. Divide this interval into n subintervals by points $x_0, x_1, \ldots, x_n$ where $a = x_0 < x_1 < \cdots < x_n = b$. This is called a *partition, net*, or *mode of subdivision* of the interval. The largest of the values $x_k - x_{k-1} = \Delta x_k$ where $k = 1, 2, \ldots, n$ is called the *norm* of the partition and is denoted by δ.

Let $M_k = $ l.u.b. $f(x)$ and $m_k = $ g.l.b. $f(x)$ in $[x_{k-1}, x_k]$ and form the sums

$$S = M_1(x_1 - x_0) + \cdots + M_n(x_n - x_{n-1}) = \sum_{k=1}^{n} M_k \Delta x_k \tag{1}$$

$$s = m_1(x_1 - x_0) + \cdots + m_n(x_n - x_{n-1}) = \sum_{k=1}^{n} m_k \Delta x_k \tag{2}$$

We call S and s the *upper* and *lower sums* respectively corresponding to the given partition. We can show that $s \leq S$ [Problem A.3].

By varying the partition, i.e. choosing different points of subdivision as well as the number of points, we obtain sets of values for S and s. Let

$I = $ g.l.b. of the values of S for all possible partitions

$J = $ l.u.b. of the values of s for all possible partitions

These values, which always exist, are called *upper* and *lower Riemann integrals* of $f(x)$ on $[a, b]$ respectively and are denoted by

$$I = \overline{\int_a^b} f(x)\, dx, \qquad J = \underline{\int_a^b} f(x)\, dx \tag{3}$$

If $I = J$ we say that $f(x)$ is *Riemann integrable* in $[a, b]$ and denote the common value by

$$\int_a^b f(x)\, dx$$

called the *Riemann definite integral* of $f(x)$ in $[a, b]$.

If $I \neq J$, then $f(x)$ is not Riemann integrable in $[a, b]$.

For a geometric interpretation see Problem A.1.

In this appendix [Appendix A], unlike the other parts of this book, all integrals will be Riemann integrals unless otherwise specified.

SOME THEOREMS ON UPPER AND LOWER SUMS AND INTEGRALS

Theorem A-1. If S, s are the upper and lower sums corresponding to a partition P and M, m are upper and lower bounds of $f(x)$ in $[a, b]$, then

$$m(b-a) \;\leq\; s \;\leq\; S \;\leq\; M(b-a)$$

Theorem A-2.   If S_1, s_1 are upper and lower sums corresponding to a new partition P_1 obtained by addition of points to the old partition P of Theorem 5-1 [often called a *refinement* of P], then

$$s \;\leqq\; s_1 \;\leqq\; S_1 \;\leqq\; S$$

i.e. upper sums do not increase and lower sums do not decrease by addition of points.

Theorem A-3.   If S_2, s_2 and S_3, s_3 are upper and lower sums corresponding to any partitions P_2 and P_3 respectively, then

$$s_3 \;\leqq\; S_2 \quad \text{or} \quad s_2 \;\leqq\; S_3$$

i.e. any lower sum is never greater than any upper sum regardless of the partition used.

Theorem A-4.   An upper integral I is never less than a lower integral J and we have $S \geqq I \geqq J \geqq s$.

NECESSARY AND SUFFICIENT CONDITION FOR RIEMANN INTEGRABILITY

Theorem A-5.   A necessary and sufficient condition for a bounded function $f(x)$ to be Riemann integrable in $[a, b]$ is that given any $\epsilon > 0$ there exists a partition with upper and lower sums S, s such that $S - s < \epsilon$.

THE RIEMANN INTEGRAL AS A LIMIT OF A SUM

The Riemann integral can also be defined as a limit of a sum. To do this choose n points of subdivision $a = x_0 < x_1 < x_2 < \cdots < x_n = b$ and also points $\xi_1, \xi_2, \ldots, \xi_n$ such that $x_{k-1} \leqq \xi_k \leqq x_k$ where $k = 1, 2, \ldots, n$. Form the sum

$$f(\xi_1)(x_1 - x_0) + f(\xi_2)(x_2 - x_1) + \cdots + f(\xi_n)(x_n - x_{n-1}) \;=\; \sum_{k=1}^{n} f(\xi_k)\, \Delta x_k$$

where $\Delta x_k = x_k - x_{k-1}$ and let the maximum of the Δx_k be equal to δ, i.e. $\max \Delta x_k = \delta$. Then we define the Riemann integral of $f(x)$ in $[a, b]$, as

$$\int_a^b f(x)\, dx \;=\; \lim_{\substack{n \to \infty \\ \delta \to 0}} \sum_{k=1}^{n} f(\xi_k) \Delta x_k \tag{4}$$

provided that the limit exists independent of the manner of choosing the points of subdivision.

It can be shown [see Problems A.8 and A.49] that this definition is equivalent to that given above.

SPECIAL TYPES OF RIEMANN INTEGRABLE FUNCTIONS

Theorem A-6.   A continuous function $f(x)$ in $[a, b]$ is Riemann integrable in $[a, b]$.

Theorem A-7.   A monotonic function $f(x)$ in $[a, b]$ is Riemann integrable in $[a, b]$.

Theorem A-8.   A function $f(x)$ of bounded variation in $[a, b]$ is Riemann integrable in $[a, b]$.

MEASURE ZERO

A set E of real numbers is said to have *Lebesgue measure zero*, or briefly *measure zero*, if given $\epsilon > 0$ there exists a countable set of open intervals I_k, $k = 1, 2, \ldots$, such that $E \subset \cup I_k$ and such that the sum of the lengths of I_k is less than ϵ, i.e. $\sum L(I_k) < \epsilon$.

A fundamental theorem on the Riemann integral is the following.

Theorem A-9.   A necessary and sufficient condition that a bounded function $f(x)$ be Riemann integrable in $[a, b]$ is that the set of discontinuities of $f(x)$ in $[a, b]$ have measure zero.

For a proof of this theorem see Problem 4.27, page 68 [compare Theorem 4-18, page 57].

THEOREMS ON RIEMANN INTEGRABLE FUNCTIONS

Theorem A-10.   If $f_1(x)$ and $f_2(x)$ are Riemann integrable in $[a, b]$, then $f_1(x) + f_2(x)$ is Riemann integrable in $[a, b]$ and

$$\int_a^b [f_1(x) + f_2(x)] \, dx \;=\; \int_a^b f_1(x) \, dx + \int_a^b f_2(x) \, dx$$

Theorem A-11.   If $f(x)$ is Riemann integrable in $[a, b]$ and c is any constant, then $c f(x)$ is Riemann integrable in $[a, b]$ and

$$\int_a^b c f(x) \, dx \;=\; c \int_a^b f(x) \, dx$$

Theorem A-12.   If $f(x)$ and $g(x)$ are Riemann integrable in $[a, b]$, then $f(x) \, g(x)$ is Riemann integrable in $[a, b]$.

Theorem A-13.   If $f(x)$ is Riemann integrable in $[a, b]$, then

$$\int_a^b f(x) \, dx \;=\; -\int_b^a f(x) \, dx, \qquad \int_a^a f(x) \, dx \;=\; 0$$

Theorem A-14.   If $f(x)$ is bounded and Riemann integrable in $[a, b]$ and c is any point of $[a, b]$, then

$$\int_a^b f(x) \, dx \;=\; \int_a^c f(x) \, dx + \int_c^b f(x) \, dx$$

Theorem A-15.   If $f(x)$ and $g(x)$ are Riemann integrable in $[a, b]$ and $f(x) \leqq g(x)$, then

$$\int_a^b f(x) \, dx \;\leqq\; \int_a^b g(x) \, dx$$

Theorem A-16.   If $f(x)$ is Riemann integrable in $[a, b]$ and has upper and lower bounds M and m in $[a, b]$ respectively, then

$$m(b - a) \;\leqq\; \int_a^b f(x) \, dx \;\leqq\; M(b - a)$$

and if μ is a number such that $m \leqq \mu \leqq M$, then

$$\int_a^b f(x) \, dx \;=\; \mu(b - a)$$

Theorem A-17 [**Mean-value theorem**].   If $f(x)$ is continuous in $[a, b]$, then there exists a number $c \in [a, b]$ such that

$$\int_a^b f(x) \, dx \;=\; (b - a) f(c)$$

Theorem A-18. If $f(x)$ is Riemann integrable in $[a, b]$ and $|f(x)| \leqq M$ for some constant M, then

$$\left| \int_a^b f(x)\, dx \right| \ \leqq \ M(b - a)$$

Theorem A-19. If $f(x)$ is Riemann integrable in $[a, b]$, then $|f(x)|$ is Riemann integrable in $[a, b]$ and

$$\left| \int_a^b f(x)\, dx \right| \ \leqq \ \int_a^b |f(x)|\, dx$$

DIFFERENTIATION AND INTEGRATION

The *indefinite Riemann integral* is defined as

$$F(x) \ = \ \int_a^x f(u)\, du \tag{5}$$

where the variable of integration has been changed to u to avoid confusion with the limit of integration x. The following theorems are important.

Theorem A-20. If $f(x)$ is bounded and Riemann integrable in $[a, b]$, then

$$F(x) \ = \ \int_a^x f(u)\, du$$

is continuous in $[a, b]$.

Theorem A-21. If $f(x)$ is Riemann integrable in $[a, b]$ and $F(x) = \int_a^x f(u)\, du$, then

$$F'(x) \ = \ \frac{d}{dx} F(x) \ = \ \frac{d}{dx} \int_a^x f(u)\, du \ = \ f(x)$$

at each point of continuity of $f(x)$.

Theorem A-22 [**Fundamental theorem of calculus**]. Let $f(x)$ be Riemann integrable in $[a, b]$ and suppose that there exists a function $F(x)$ continuous on $[a, b]$ such that $F'(x) = f(x)$. Then

$$\int_a^b F'(x)\, dx \ = \ \int_a^b f(x)\, dx \ = \ F(b) - F(a)$$

or

$$\int_a^x F'(u)\, du \ = \ \int_a^x f(u)\, du \ = \ F(x) - F(a)$$

Theorem A-23 [**Integration by parts**]. Let $f(x)$ and $g(x)$ be Riemann integrable in $[a, b]$ and suppose that $F(x)$ and $G(x)$ are such that $F'(x) = f(x)$ and $G'(x) = g(x)$ in $[a, b]$. Then

$$\int_a^b F(x)\, g(x)\, dx \ = \ F(b)\, G(b) - F(a)\, G(a) - \int_a^b f(x)\, G(x)\, dx$$

$$= \ F(x)\, G(x) \Big|_a^b - \int_a^b f(x)\, G(x)\, dx$$

Theorem A-24 [**Change of variables**]. Let $f(x)$ be continuous in $[a, b]$. Let $x = g(u)$ have a continuous derivative in $[\alpha, \beta]$ where $a = g(\alpha)$, $b = g(\beta)$. Then

$$\int_a^b f(x)\, dx \ = \ \int_\alpha^\beta f[g(u)]\, g'(u)\, du$$

THEOREMS ON SEQUENCES AND SERIES

Theorem A-25. Let $\langle f_n(x) \rangle$ be a sequence of continuous functions which converges uniformly in $[a, b]$ to a function $f(x)$. Then

$$\lim_{n \to \infty} \int_a^b f_n(x) \, dx \;=\; \int_a^b f(x) \, dx \;=\; \int_a^b \lim_{n \to \infty} f_n(x) \, dx$$

Theorem A-26. Let $\sum_{n=1}^{\infty} u_n(x)$ be a uniformly convergent series of continuous functions in $[a, b]$ and let $s(x)$ be the sum of the series. Then

$$\int_a^b s(x) \, dx \;=\; \int_a^b \left[\sum_{n=1}^{\infty} u_n(x) \right] dx \;=\; \sum_{n=1}^{\infty} \int_a^b u_n(x) \, dx$$

This is a reformulation of Theorem A-25 using series.

Theorem A-27. Let $\sum_{n=1}^{\infty} u_n(x)$ converge to $s(x)$ in $[a, b]$. Suppose that the derivatives $u_n'(x)$ exist and are continuous in $[a, b]$ and that $\sum_{n=1}^{\infty} u_n'(x)$ is uniformly convergent in $[a, b]$. Then

$$s'(x) \;=\; \frac{d}{dx} \sum_{n=1}^{\infty} u_n(x) \;=\; \sum_{n=1}^{\infty} u_n'(x)$$

IMPROPER RIEMANN INTEGRALS

If the interval $[a, b]$ is infinite or if $f(x)$ becomes infinite at one or more points of $[a, b]$, then the Riemann integral of $f(x)$ is often referred to as an *improper integral*. Such integrals are defined by appropriate limiting procedures and are said to exist if the corresponding limits exist. The following are examples of such definition.

$$\int_a^{\infty} f(x) \, dx \;=\; \lim_{b \to \infty} \int_a^b f(x) \, dx \tag{6}$$

$$\int_{-\infty}^b f(x) \, dx \;=\; \lim_{a \to -\infty} \int_a^b f(x) \, dx \tag{7}$$

$$\int_{-\infty}^{\infty} f(x) \, dx \;=\; \lim_{\substack{a \to -\infty \\ b \to \infty}} \int_a^b f(x) \, dx \tag{8}$$

If $f(x)$ becomes unbounded when $x = c$, then we define

$$\int_a^b f(x) \, dx \;=\; \lim_{\substack{\epsilon_1 \to 0 \\ \epsilon_2 \to 0}} \left[\int_a^{c - \epsilon_1} f(x) \, dx + \int_{c + \epsilon_2}^b f(x) \, dx \right] \tag{9}$$

If the limit in (9) does not exist for $\epsilon_1 \neq \epsilon_2$ but does exist when $\epsilon_1 = \epsilon_2$, we say that the corresponding value of the limit is the *Cauchy principal value* of $\int_a^b f(x) \, dx$. Also in such case we say that the Riemann integral of $f(x)$ exists in the *Cauchy principal value sense*. A similar remark can be made for (8) for which the Cauchy principal value, if it exists, is obtained by taking $a = -b$ and letting $b \to \infty$.

Solved Problems

DEFINITION OF THE RIEMANN INTEGRAL

A.1. Interpret geometrically the case where $f(x)$ is Riemann integrable in $[a, b]$.

Refer to Fig. A-1 which shows the curve C representing $y = f(x)$. The lower sum s corresponding to the partition shown represents the sum of areas of the shaded rectangles while the upper sum S represents the sum of areas of the larger rectangles. Clearly $s \leqq S$, from the figure.

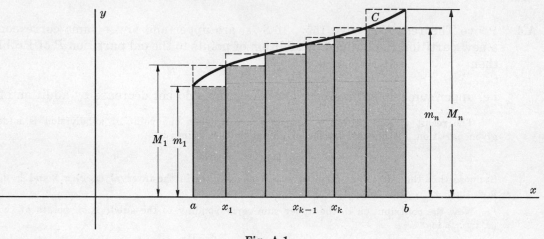

Fig. A-1

The value $\displaystyle\int_a^b f(x)\, dx$ in this case represents the area of the region bounded by the curve, the x axis and the ordinates $x = a$ and $x = b$.

A.2. Prove that the function $f(x) = \begin{cases} 1, & x \text{ rational} \\ 0, & x \text{ irrational} \end{cases}$ defined in the interval $[a, b]$ is not Riemann integrable.

In any subinterval $[x_{k-1}, x_k]$ of $[a, b]$ we have $M_k = 1$, $m_k = 0$ since the rational and irrational numbers are dense in any subinterval. Then if S, s are the upper and lower sums corresponding to any partition, we have

$$S = \sum_{k=1}^n M_k\, \Delta x_k = \sum_{k=1}^n 1 \cdot \Delta x_k = b - a$$

$$s = \sum_{k=1}^n m_k\, \Delta x_k = \sum_{k=1}^n 0 \cdot \Delta x_k = 0$$

Thus $I = b - a$, $J = 0$ so that $I \neq J$ and $f(x)$ is not Riemann integrable.

THEOREMS ON UPPER AND LOWER SUMS AND INTEGRALS

A.3. Prove Theorem A-1, page 154: If S, s are the upper and lower sums corresponding to a partition P and M, m are upper and lower bounds of $f(x)$ in $[a, b]$, then

$$m(b-a) \leqq s \leqq S \leqq M(b-a)$$

From equations (1) and (2), page 154, we have

$$S = \sum_{k=1}^n M_k\, \Delta x_k, \qquad s = \sum_{k=1}^n m_k\, \Delta x_k$$

Then since $m \leqq m_k \leqq M_k \leqq M$ we have on multiplying by Δx_k and summing over k from 1 to n,

$$\sum_{k=1}^{n} m \, \Delta x_k \;\leqq\; \sum_{k=1}^{n} m_k \, \Delta x_k \;\leqq\; \sum_{k=1}^{n} M_k \, \Delta x_k \;\leqq\; \sum_{k=1}^{n} M \, \Delta x_k$$

i.e.

$$m \sum_{k=1}^{n} \Delta x_k \;\leqq\; s \;\leqq\; S \;\leqq\; M \sum \Delta x_k$$

or

$$m(b-a) \;\leqq\; s \;\leqq\; S \;\leqq\; M(b-a)$$

A.4. Prove Theorem A-2, page 155: If S_1, s_1 are upper and lower sums corresponding to a new partition P_1 obtained by addition of points to the old partition P of Problem A.3, then

$$s \leqq s_1 \leqq S_1 \leqq S$$

i.e. upper sums do not increase and lower sums do not decrease by addition of points.

The result will be proved if we can prove it when one point of subdivision is added to the given partition. To do this let the given subdivision points be

$$a \;=\; x_0 < x_1 < x_2 < \cdots < x_n \;=\; b$$

Suppose that the additional point of subdivision occurs in the interval (x_{p-1}, x_p) and is denoted by u so that $x_{p-1} < u < x_p$.

Now the contribution to the upper sum corresponding to the subdivision points of the interval (x_{p-1}, x_p) is

$$M_p(x_p - x_{p-1}) \tag{1}$$

If we denote the least upper bounds of $f(x)$ in (x_{p-1}, u) and (u, x_p) by $M_p^{(1)}$ and $M_p^{(2)}$ respectively, then the contribution to the upper sum when the additional point of subdivision is taken into account is

$$M_p^{(1)}(u - x_{p-1}) \;+\; M_p^{(2)}(x_p - u) \tag{2}$$

The change in the original upper sum caused by the additional subdivision point is given by the difference between (1) and (2), i.e.

$$(M_p^{(1)} - M_p)(u - x_{p-1}) \;+\; (M_p^{(2)} - M_p)(x_p - u)$$

Since this is negative or zero [because $M_p^{(1)} \leqq M_p$, $M_p^{(2)} \leqq M_p$, and $u > x_{p-1}$, $x_p > u$], it follows that the upper sum cannot increase by adding a point of subdivision and the required result is proved.

Similarly we can prove that the lower sums cannot decrease by adding points of subdivision [see Problem A.38].

A.5. Prove Theorem A-3, page 155: If S_2, s_2 and S_3, s_3 are upper and lower sums corresponding to any partitions P_2 and P_3 respectively, then

$$s_3 \leqq S_2 \quad \text{or} \quad s_2 \leqq S_3$$

i.e. any lower sum is never greater than any upper sum regardless of the partition used.

The proof is identical with that given in Problem 4.5, page 59.

A.6. Prove Theorem A-4, page 155: An upper integral I is never less than a lower integral J and we have $S \geqq I \geqq J \geqq s$.

The proof is identical with that given in Problem 4.6, page 59.

A.7. Prove Theorem A-5, page 155: A necessary and sufficient condition for a bounded function $f(x)$ to be Riemann integrable in $[a, b]$ is that given any $\epsilon > 0$, there exists a partition with upper and lower sums S, s such that $S - s < \epsilon$.

The proof is identical with that of Problem 4.8, page 59, except that the words "Riemann integrable" replace "Lebesgue integrable".

THE RIEMANN INTEGRAL AS THE LIMIT OF A SUM

A.8. Prove that the definition of a Riemann integral as a limit of a sum [page 155] follows from the definition on page 154.

We must show that if $\epsilon > 0$, then we can choose $\delta > 0$ such that if $x_{k-1} \leqq \xi_k \leqq x_k$,

$$\left| \sum_{k=1}^{n} f(\xi_k) \, \Delta x_k - \int_a^b f(x) \, dx \right| \; < \; \epsilon$$

whenever
$$\Delta x_k \; \leqq \; \delta$$

Let S, s be the upper and lower sums corresponding to the given partition. Then since $m_k \leqq f(\xi_k) \leqq M_k$ where M_k, m_k are the l.u.b. and g.l.b. of $f(x)$ in (x_{k-1}, x_k), we have

$$s \; \leqq \; \sum_{k=1}^{n} f(\xi_k) \, \Delta x_k \; \leqq \; S \tag{1}$$

Also we have
$$S \; \geqq \; \int_a^b f(x) \, dx \; \geqq \; s \tag{2}$$

or
$$-S \; \leqq \; -\int_a^b f(x) \, dx \; \leqq \; -s \tag{3}$$

Adding (1) and (3),

$$-(S - s) \; \leqq \; \sum_{k=1}^{n} f(\xi_k) \, \Delta x_k - \int_a^b f(x) \, dx \; \leqq \; S - s$$

or since $S - s > 0$,
$$\left| \sum_{k=1}^{n} f(\xi_k) \, \Delta x_k - \int_a^b f(x) \, dx \right| \; \leqq \; S - s$$

Now since we can choose the largest values of Δx_k [i.e. the norm] so small that $S - s < \epsilon$, the required result follows.

We can also prove that the definition of the Riemann integral given on page 154 follows from the definition as a limit of a sum [see Problem A.49]. Thus the two definitions are equivalent.

THEOREMS INVOLVING THE RIEMANN INTEGRAL

A.9. Prove Theorem A-6, page 155: A continuous function $f(x)$ in $[a, b]$ is Riemann integrable in $[a, b]$.

Since $f(x)$ is continuous in the closed interval $[a, b]$, it is uniformly continuous. Let $x^{(1)}$ and $x^{(2)}$ be any two points of an interval (x_{k-1}, x_k). Then given $\epsilon > 0$, there exists $\delta > 0$ such that

$$|f(x^{(1)}) - f(x^{(2)})| \; < \; \frac{\epsilon}{b - a} \quad \text{whenever} \quad |x^{(1)} - x^{(2)}| \; < \; \delta$$

Thus we can choose points of subdivision so that

$$M_k - m_k \; < \; \frac{\epsilon}{b - a}$$

If the upper and lower sums corresponding to this partition are given by

$$S \; = \; \sum_{k=1}^{n} M_k \, \Delta x_k, \quad s \; = \; \sum_{k=1}^{n} m_k \, \Delta x_k$$

then we have

$$S - s = \sum_{k=1}^{n} (M_k - m_k) \Delta x_k < \sum_{k=1}^{n} \frac{\epsilon}{b-a} \Delta x_k = \epsilon$$

Thus $S - s < \epsilon$ and it follows by Problem A.7 that $f(x)$ is Riemann integrable.

A.10. Prove Theorem A-7, page 155: A monotonic function $f(x)$ in $[a, b]$ is Riemann integrable in $[a, b]$.

We shall assume that $f(x)$ is monotonic increasing. The case where $f(x)$ is monotonic decreasing can be proved similarly [or by considering $-f(x)$ instead of $f(x)$].

By definition we have for the given partition $a = x_0 < x_1 < \cdots < x_n = b$,

$$f(a) = f(x_0) \leqq f(x_1) \leqq \cdots \leqq f(x_n) = f(b)$$

Then it is clear that $m_k = f(x_{k-1})$, $M_k = f(x_k)$ so that

$$S = \sum_{k=1}^{n} f(x_k) \Delta x_k, \qquad s = \sum_{k=1}^{n} f(x_{k-1}) \Delta x_k$$

or

$$S - s = \sum_{k=1}^{n} [f(x_k) - f(x_{k-1})] \Delta x_k \qquad (1)$$

If we choose the partition so that assuming $f(b) \neq f(a)$

$$\Delta x_k < \frac{\epsilon}{f(b) - f(a)} \qquad (2)$$

we have from (1), since $f(x_k) \geqq f(x_{k-1})$,

$$S - s < \frac{\epsilon}{f(b) - f(a)} \sum_{k=1}^{n} [f(x_k) - f(x_{k-1})]$$

$$= \frac{\epsilon}{f(b) - f(a)} [f(b) - f(a)] = \epsilon$$

Thus by Problem A.7 the required result follows.

A.11. Prove Theorem A-8, page 155: A function $f(x)$ of bounded variation in $[a, b]$ is Riemann integrable in $[a, b]$.

This follows at once from Problem A.10 and the fact that a function of bounded variation can be expressed as the difference of two monotonic increasing functions.

A.12. Prove Theorem A-10, page 156: If $f_1(x)$, and $f_2(x)$, are Riemann integrable in $[a, b]$, then $f_1(x) + f_2(x)$ is Riemann integrable in $[a, b]$ and

$$\int_a^b [f_1(x) + f_2(x)] \, dx = \int_a^b f_1(x) \, dx + \int_a^b f_2(x) \, dx$$

Let $M_k^{(1)}, m_k^{(1)}$; $M_k^{(2)}, m_k^{(2)}$; M_k, m_k, be the least upper bounds and greatest lower bounds in (x_{k-1}, x_k) of $f_1(x)$; $f_2(x)$; $f_1(x) + f_2(x)$ respectively. Then we have

$$M_k \leqq M_k^{(1)} + M_k^{(2)}, \qquad m_k \geqq m_k^{(1)} + m_k^{(2)} \qquad (1)$$

Let us call $S^{(1)}, s^{(1)}$; $S^{(2)}, s^{(2)}$; S, s, the upper and lower sums corresponding to $f_1(x)$; $f_2(x)$; $f_1(x) + f_2(x)$ respectively. Then using (1), we find

$$S \leqq S^{(1)} + S^{(2)}, \qquad s \geqq s^{(1)} + s^{(2)} \qquad (2)$$

Using the symbols $I^{(1)}$; $I^{(2)}$; I for the corresponding upper integrals and $J^{(1)}$; $J^{(2)}$; J for lower integrals, we have from (2)

$$I \leq I^{(1)} + I^{(2)}, \qquad J \geq J^{(1)} + J^{(2)} \tag{3}$$

But since $f_1(x)$ and $f_2(x)$ are Riemann integrable, we have $I^{(1)} = J^{(1)}$, $I^{(2)} = J^{(2)}$. Thus

$$I \leq J^{(1)} + J^{(2)}, \qquad J \geq J^{(1)} + J^{(2)} \tag{4}$$

so that $I \leq J$. But from Problem A.6, $I \geq J$ and so it follows that $I = J = I^{(1)} + I^{(2)} = J^{(1)} + J^{(2)}$.

Thus $f_1(x) + f_2(x)$ is Riemann integrable and

$$\int_a^b [f_1(x) + f_2(x)] \, dx = \int_a^b f_1(x) \, dx + \int_a^b f_2(x) \, dx$$

Another Method, using limit of a sum definition.

If $\max \Delta x_k = \delta$ we have, using the notation of page 155,

$$\int_a^b [f_1(x) + f_2(x)] \, dx = \lim_{\substack{n \to \infty \\ \delta \to 0}} \sum_{k=1}^n [f_1(\xi_k) + f_2(\xi_k)] \, \Delta x_k$$

$$= \lim_{\substack{n \to \infty \\ \delta \to 0}} \sum_{k=1}^n f_1(\xi_k) \, \Delta x_k + \lim_{\substack{n \to \infty \\ \delta \to 0}} \sum_{k=1}^n f_2(\xi_k) \, \Delta x_k$$

$$= \int_a^b f_1(x) \, dx + \int_a^b f_2(x) \, dx$$

A.13. Give an example to show that $|f(x)|$ can be Riemann integrable even though $f(x)$ is not Riemann integrable.

Consider $f(x) = \begin{cases} 1 & \text{if } x \in [a, b] \text{ is rational} \\ -1 & \text{if } x \in [a, b] \text{ is irrational} \end{cases}$. Then as in Problem A.2 we can show that $f(x)$ is not Riemann integrable in $[a, b]$.

However $|f(x)| = 1$ is Riemann integrable in $[a, b]$ as shown in Problem A.34.

A.14. Prove Theorem A-14, page 156: If $f(x)$ is bounded and Riemann integrable in $[a, b]$ and c is any point of $[a, b]$, then

$$\int_a^b f(x) \, dx = \int_a^c f(x) \, dx + \int_c^b f(x) \, dx$$

Suppose that c is not a point of subdivision in defining upper and lower sums. By Problem A.4 if we add the point c to the partition, the upper sum S is not increased.

Denoting the upper sums for the intervals (a, c) and (c, b) by S_1 and S_2, we have

$$S \geq S_1 + S_2 \geq \int_a^c f(x) \, dx + \int_c^b f(x) \, dx \tag{1}$$

In a similar way we find for the corresponding lower sums

$$s \leq s_1 + s_2 \leq \int_a^c f(x) \, dx + \int_c^b f(x) \, dx \tag{2}$$

Thus if I and J are the upper and lower integrals corresponding to $f(x)$, we have

$$S \geq I \geq \int_a^c f(x) \, dx + \int_c^b f(x) \, dx \geq J \geq s \tag{3}$$

Now since $f(x)$ is Riemann integrable in $[a, b]$, we have

$$I = J = \int_a^b f(x) \, dx \tag{4}$$

and thus from (3) and (4),

$$\int_a^b f(x) \, dx = \int_a^c f(x) \, dx + \int_c^b f(x) \, dx$$

A.15. Prove Theorem A-15, page 156: If $f(x)$ and $g(x)$ are Riemann integrable in $[a, b]$ and $f(x) \leqq g(x)$, then

$$\int_a^b f(x)\, dx \;\; \leqq \;\; \int_a^b g(x)\, dx$$

Let $M_k^{(1)}, m_k^{(1)}$ and $M_k^{(2)}, m_k^{(2)}$ be the least upper bounds and greatest lower bounds in (x_{k-1}, x_k) corresponding to $f(x)$ and $g(x)$ respectively. Also let $S^{(1)}, s^{(1)}$ and $S^{(2)}, s^{(2)}$ be the corresponding upper and lower sums. Then we have, since $f(x) \leqq g(x)$,

$$M_k^{(1)} \;\leqq\; M_k^{(2)}, \qquad m_k^{(1)} \;\leqq\; m_k^{(2)} \tag{1}$$

so that

$$S^{(1)} \;\leqq\; S^{(2)}, \qquad s^{(1)} \;\leqq\; s^{(2)} \tag{2}$$

If we now let the upper and lower integrals be $I^{(1)}, J^{(1)}$ and $I^{(2)}, J^{(2)}$ corresponding to $f(x)$ and $g(x)$ respectively, we have

$$I^{(1)} \;\leqq\; I^{(2)}, \qquad J^{(1)} \;\leqq\; J^{(2)} \tag{3}$$

Then since

$$I^{(1)} \,=\, J^{(1)} \,=\, \int_a^b f(x)\, dx, \qquad I^{(2)} \,=\, J^{(2)} \,=\, \int_a^b g(x)\, dx \tag{4}$$

we have

$$\int_a^b f(x)\, dx \;\; \leqq \;\; \int_a^b g(x)\, dx$$

A.16. Prove Theorem A-16, page 156: If $f(x)$ is Riemann integrable in $[a, b]$ and has upper and lower bounds M and m in $[a, b]$ respectively, then

$$m(b-a) \;\; \leqq \;\; \int_a^b f(x)\, dx \;\; \leqq \;\; M(b-a)$$

and if μ is a number such that $m \leqq \mu \leqq M$, then

$$\int_a^b f(x)\, dx \;\; = \;\; \mu(b-a)$$

We have

$$m \;\leqq\; f(x) \;\leqq\; M$$

Then by integrating from a to b using Problem A.15, we obtain

$$\int_a^b m\, dx \;\; \leqq \;\; \int_a^b f(x)\, dx \;\; \leqq \;\; \int_a^b M\, dx$$

or

$$m(b-a) \;\; \leqq \;\; \int_a^b f(x)\, dx \;\; \leqq \;\; M(b-a) \tag{1}$$

From this it follows that there is a number μ between m and M such that $m(b-a) \leqq \mu(b-a) \leqq M(b-a)$ and

$$\int_a^b f(x)\, dx \;\; = \;\; \mu(b-a) \tag{2}$$

A.17. Prove Theorem A-17 [Mean-value theorem], page 156: If $f(x)$ is continuous in $[a, b]$, then there exists a number $c \in [a, b]$ such that

$$\int_a^b f(x)\, dx \;\; = \;\; (b-a)f(c)$$

Since $f(x)$ is continuous in $[a, b]$, it is Riemann integrable in $[a, b]$ by Problem A.9. Thus by Problem A.16 there is a number μ between the minimum and maximum values m and M of $f(x)$ such that

$$\int_a^b f(x)\, dx \;\; = \;\; \mu(b-a) \tag{1}$$

But a continuous function takes on all values between its minimum and maximum values and so there is a number c such that $f(c) = \mu$ [see Theorem 1-25, page 8]. Using this in (1) yields the required result.

A.18. Prove Theorem A-19, page 157: If $f(x)$ is Riemann integrable in $[a, b]$, then   (a) $|f(x)|$ is Riemann integrable in $[a, b]$ and (b) $\left| \int_a^b f(x)\,dx \right| \leqq \int_a^b |f(x)|\,dx$.

(a) Let

$$F_1(x) \;=\; \begin{cases} f(x), & f(x) \geqq 0 \\ 0, & \text{otherwise} \end{cases}, \qquad F_2(x) \;=\; \begin{cases} -f(x), & f(x) \leqq 0 \\ 0, & \text{otherwise} \end{cases}$$

Then we have

$$f(x) \;=\; F_1(x) - F_2(x), \qquad |f(x)| \;=\; F_1(x) + F_2(x) \tag{1}$$

Now since $f(x)$ is Riemann integrable, so also are $F_1(x)$ and $F_2(x)$. Thus by Problem A.12, $F_1(x) + F_2(x) = |f(x)|$ is Riemann integrable.

(b) From (1) we have

$$\int_a^b f(x)\,dx \;=\; \int_a^b F_1(x)\,dx \;-\; \int_a^b F_2(x)\,dx$$

$$\int_a^b |f(x)|\,dx \;=\; \int_a^b F_1(x)\,dx \;+\; \int_a^b F_2(x)\,dx$$

Then

$$\left| \int_a^b f(x)\,dx \right| \;\leqq\; \left| \int_a^b F_1(x)\,dx \right| + \left| \int_a^b F_2(x)\,dx \right|$$

$$=\; \int_a^b F_1(x)\,dx \;+\; \int_a^b F_2(x)\,dx \;=\; \int_a^b |f(x)|\,dx$$

DIFFERENTIATION AND INTEGRATION

A.19. Prove Theorem A-20, page 157: If $f(x)$ is bounded and Riemann integrable in $[a, b]$, then

$$F(x) \;=\; \int_a^x f(u)\,du$$

is continuous in $[a, b]$.

Let h be chosen so that $x + h \in [a, b]$. Then

$$F(x + h) - F(x) \;=\; \int_a^{x+h} f(u)\,du \;-\; \int_a^x f(u)\,du$$

$$=\; \int_x^{x+h} f(u)\,du \;=\; \mu h$$

by Problem A.16.

Thus if $|h| < \delta$, we have

$$|F(x + h) - F(x)| \;=\; |\mu h| \;=\; |\mu|\,|h| \;<\; |\mu|\,\delta \;=\; \epsilon$$

i.e. $|F(x + h) - F(x)| \;<\; \epsilon$ for $|h| < \delta$

and so $F(x)$ is continuous in $[a, b]$.

A.20. Prove Theorem A-21, page 157: If $f(x)$ is Riemann integrable in $[a, b]$ and $F(x) = \int_a^x f(u)\,du$, then

$$F'(x) \;=\; \frac{d}{dx} F(x) \;=\; \frac{d}{dx} \int_a^x f(u)\,du \;=\; f(x)$$

at each point of continuity of $f(x)$.

Suppose that x_0 is a point of continuity of f. Then given $\epsilon > 0$, there exists $\delta > 0$ such that

$$|f(x) - f(x_0)| \;<\; \epsilon \qquad \text{whenever} \qquad |x - x_0| \;<\; \delta \tag{1}$$

Now
$$F(x_0 + h) - F(x_0) = \int_a^{x_0+h} f(u)\,du - \int_a^{x_0} f(u)\,du = \int_{x_0}^{x_0+h} f(u)\,du$$

Thus
$$\frac{F(x_0+h) - F(x_0)}{h} - f(x_0) = \frac{1}{h}\int_{x_0}^{x_0+h} f(u)\,du - \frac{1}{h}\int_{x_0}^{x_0+h} f(x_0)\,du$$

$$= \frac{1}{h}\int_{x_0}^{x_0+h} [f(u) - f(x_0)]\,du$$

and it follows from (1) that if $0 < |h| < \delta$,

$$\left| \frac{F(x_0+h) - F(x_0)}{h} - f(x_0) \right| = \frac{1}{|h|}\left| \int_{x_0}^{x_0+h} [f(u) - f(x_0)]\,du \right|$$

$$\leq \frac{1}{|h|}\left| \int_{x_0}^{x_0+h} |f(u) - f(x_0)|\,du \right|$$

$$< \epsilon$$

Thus
$$\lim_{h \to 0} \frac{F(x_0+h) - F(x_0)}{h} = f(x_0)$$

or
$$F'(x_0) = f(x_0)$$

i.e. $F'(x) = f(x)$ at each point of continuity of $f(x)$.

A.21. (a) Prove *Rolle's theorem*: Let $f(x)$ be continuous in $[a, b]$ and suppose that $f(a) = f(b) = 0$. Then if $f'(x)$ exists in (a, b), there is a point c in (a, b) such that $f'(c) = 0$.

(b) Prove the *law of the mean* or *mean-value theorem* for derivatives: Let $f(x)$ be continuous in $[a, b]$ and suppose that $f'(x)$ exists in (a, b). Then there is a point c in (a, b) such that

$$f'(c) = \frac{f(b) - f(a)}{b - a}$$

(a) The theorem is trivial if $f(x) = 0$ throughout $[a, b]$. Suppose then that $f(x) > 0$ at one point of (a, b). Then there is a point c in (a, b) at which $f(x)$ attains a maximum value [see Theorem 1-25, page 8]. Now if $f'(c) > 0$, i.e. if

$$\lim_{h \to 0} \frac{f(c + h) - f(c)}{h} > 0 \tag{1}$$

then

$$\frac{f(c + h) - f(c)}{h} > 0 \quad \text{for } |h| < \delta \tag{2}$$

Thus if $h > 0$, it follows from (2) that $f(c + h) > f(c)$ so that $f(x)$ does not have a maximum at c and we have a contradiction. Thus $f'(c)$ cannot be greater than zero. Similarly we can show that $f'(c)$ cannot be less than zero and so we must have $f'(c) = 0$.

The same result follows if we assume that $f(x) < 0$ at some point of (a, b).

(b) Consider the function

$$G(x) = f(x) - f(a) - \left[\frac{f(b) - f(a)}{b - a}\right](x - a) \tag{3}$$

Then if $f(x)$ is continuous in $[a, b]$, so is $G(x)$. Also if $f'(x)$ exists in (a, b), then $G'(x)$ exists in (a, b). Furthermore we have $G(a) = G(b) = 0$. Thus $G(x)$ satisfies Rolle's theorem of part (a) and so there exists a point c in (a, b) such that $G'(c) = 0$. But this means that

$$G'(c) = f'(c) - \left[\frac{f(b) - f(a)}{b - a}\right] = 0 \tag{4}$$

and the required result follows.

A.22. Prove Theorem A-22 [Fundamental theorem of calculus], page 157:   Let $f(x)$ be Riemann integrable in $[a, b]$ and suppose that there exists a function $F(x)$ continuous on $[a, b]$ such that $F'(x) = f(x)$.   Then

$$\int_a^b f(x)\, dx \;=\; F(b) - F(a) \qquad \text{or} \qquad \int_a^x f(u)\, du \;=\; F(x) - F(a)$$

Suppose that we form any partition $a = x_0 < x_1 < \cdots < x_n = b$ of the interval $[a, b]$. By the mean-value theorem of Problem A.21, we have [since $F(x)$ is continuous in $[a, b]$ and has a derivative in $a < x < b$]

$$F(x_k) - F(x_{k-1}) \;=\; (x_k - x_{k-1})F'(c_k) \qquad x_{k-1} < c_k < x_k$$

i.e.
$$F(x_k) - F(x_{k-1}) \;=\; (x_k - x_{k-1})f(c_k)$$

Summing over k from $k = 1$ to n, we have

$$\sum_{k=1}^n [F(x_k) - F(x_{k-1})] \;=\; F(b) - F(a) \;=\; \sum_{k=1}^n (x_k - x_{k-1})f(c_k)$$

Then taking the limit as $n \to \infty$ and $\max (x_k - x_{k-1}) \to 0$, it follows that

$$\int_a^b f(x)\, dx \;=\; F(b) - F(a) \qquad \text{or} \qquad \int_a^x f(u)\, du \;=\; F(x) - F(a)$$

A.23. (a) Prove that the derivative of $\;F(x) = \begin{cases} x^2 \sin (1/x), & x \neq 0 \\ 0, & x = 0 \end{cases}\;$ is given by

$$F'(x) \;=\; \begin{cases} 2x \sin (1/x) - \cos (1/x), & x \neq 0 \\ 0, & x = 0 \end{cases}$$

(b) Show that
$$\int_{-2/\pi}^{2/\pi} F'(x)\, dx \;=\; \frac{8}{\pi^2}$$

(a) If $x \neq 0$, we can use the usual rules of differential calculus to obtain

$$\frac{d}{dx}\{x^2 \sin (1/x)\} \;=\; x^2 \left\{\cos \frac{1}{x}\right\}\left(-\frac{1}{x^2}\right) + 2x \sin \frac{1}{x} \;=\; 2x \sin \frac{1}{x} - \cos \frac{1}{x}$$

If $x = 0$, the derivative is given by

$$\lim_{h \to 0} \frac{F(h) - F(0)}{h} \;=\; \lim_{h \to 0} \frac{h^2 \sin (1/h) - 0}{h} \;=\; \lim_{h \to 0} h \sin \frac{1}{h} \;=\; 0$$

(b) By part (a), $F'(x)$ exists at all values of x in $(-2/\pi, 2/\pi)$.   Thus by Problem A.22,

$$\int_{-2/\pi}^{2/\pi} F'(x)\, dx \;=\; F(2/\pi) - F(-2/\pi) \;=\; \frac{8}{\pi^2}$$

A.24. Prove Theorem A-23 [Integration by parts], page 157:   Let $f(x)$ and $g(x)$ be Riemann integrable in $[a, b]$ and suppose that $F(x)$ and $G(x)$ are such that $F'(x) = f(x)$ and $G'(x) = g(x)$ in $[a, b]$.   Then

$$\int_a^b F(x)\, g(x)\, dx \;=\; F(b)\, G(b) - F(a)\, G(a) - \int_a^b f(x)\, G(x)\, dx$$

$$=\; F(x)\, G(x)\Big|_a^b \;-\; \int_a^b f(x)\, G(x)\, dx$$

Since $F(x)$ and $G(x)$ have derivatives in $[a, b]$, they are continuous and thus Riemann integrable in $[a, b]$.   Then since $F'(x)$ and $G'(x)$ are also Riemann integrable in $[a, b]$, it follows that $F(x)\, G'(x)$ and $F'(x)\, G(x)$ and thus their sum $F(x)\, G'(x) + F'(x)\, G(x) = \dfrac{d}{dx}[F(x)\, G(x)]$ is integrable in $[a, b]$. Then from Problem A.22,

$$\int_a^b \frac{d}{dx}[F(x)\, G(x)]\, dx \;=\; F(x)\, G(x)\Big|_a^b \;=\; F(b)\, G(b) - F(a)\, G(a)$$

or
$$\int_a^b [F(x)\, G'(x) + F'(x)\, G(x)]\, dx \;=\; F(b)\, G(b) - F(a)\, G(a)$$

i.e.
$$\int_a^b F(x)\, g(x)\, dx \;=\; F(b)\, G(b) - F(a)\, G(a) - \int_a^b f(x)\, G(x)\, dx$$

since $F'(x) = f(x),\; G'(x) = g(x)$.

THEOREMS ON SEQUENCES AND SERIES

A.25. Prove Theorem A-26, page 158. Let $\displaystyle\sum_{n=1}^{\infty} u_n(x)$ be a uniformly convergent series of continuous functions in $[a, b]$ and let $s(x)$ be the sum of the series. Then

$$\int_a^b s(x)\, dx \;=\; \int_a^b \left[\sum_{n=1}^{\infty} u_n(x) \right] dx \;=\; \sum_{n=1}^{\infty} \int_a^b u_n(x)\, dx$$

If a function is continuous in $[a, b]$, its integral exists [see Problem A.9]. Then since $s(x), s_n(x)$ and $r_n(x)$ are continuous [see Problem 1.53, page 23], we have

$$\int_a^b s(x)\, dx \;=\; \int_a^b s_n(x)\, dx + \int_a^b r_n(x)\, dx$$

where $s_n(x) = u_1(x) + \cdots + u_n(x),\quad r_n(x) = u_{n+1}(x) + u_{n+2}(x) + \cdots$.

To prove the theorem we must show that

$$\left| \int_a^b s(x)\, dx - \int_a^b s_n(x)\, dx \right| \;=\; \left| \int_a^b r_n(x)\, dx \right|$$

can be made arbitrarily small by choosing n large enough. This however follows at once, since by the uniform convergence of the series, given any $\epsilon > 0$ there exists n_0 such that $|r_n(x)| < \dfrac{\epsilon}{b-a}$ for $n > n_0$ independent of x in $[a, b]$ and so

$$\left| \int_a^b r_n(x)\, dx \right| \;\leq\; \int_a^b |r_n(x)|\, dx \;<\; \int_a^b \frac{\epsilon}{b-a}\, dx \;=\; \epsilon$$

This is equivalent to the statements

$$\int_a^b s(x)\, dx \;=\; \lim_{n \to \infty} \int_a^b s_n(x)\, dx \quad \text{or} \quad \lim_{n \to \infty} \int_a^b s_n(x)\, dx \;=\; \int_a^b \left\{ \lim_{n \to \infty} s_n(x) \right\} dx$$

A.26. Prove Theorem A-27, page 158: Let $\displaystyle\sum_{n=1}^{\infty} u_n(x)$ converge to $s(x)$ in $[a, b]$. Suppose that the derivatives $u_n'(x)$ exist and are continuous in $[a, b]$ and that $\displaystyle\sum_{n=1}^{\infty} u_n'(x)$ is uniformly convergent in $[a, b]$. Then

$$s'(x) \;=\; \frac{d}{dx} \sum_{n=1}^{\infty} u_n(x) \;=\; \sum_{n=1}^{\infty} u_n'(x)$$

Let $g(x) = \displaystyle\sum_{n=1}^{\infty} u_n'(x)$. By hypothesis this series converges uniformly in $[a, b]$ so that by Problem A.25 we can integrate term by term from a to x, where $x \in [a, b]$, to obtain

$$\int_a^x g(t)\, dt \;=\; \sum_{n=1}^{\infty} \int_a^x u_n'(t)\, dt \;=\; \sum_{n=1}^{\infty} [u_n(x) - u_n(a)]$$

$$=\; \sum_{n=1}^{\infty} u_n(x) - \sum_{n=1}^{\infty} u_n(a) \;=\; s(x) - s(a)$$

Differentiating both sides of $\displaystyle\int_a^x g(t)\, dt = s(x) - s(a)$ then shows that $g(x) = s'(x)$, which proves the theorem.

A.27. Let $\ s_n(x) = nxe^{-nx^2}, \ n = 1, 2, 3, \ldots, \ 0 \leqq x \leqq 1.$

(a) Determine whether $\ \displaystyle\lim_{n \to \infty} \int_0^1 s_n(x)\, dx \ = \ \int_0^1 \lim_{n \to \infty} s_n(x)\, dx.$

(b) Explain the result in (a).

(a) $\displaystyle \int_0^1 s_n(x)\, dx \ = \ \int_0^1 nxe^{-nx^2}\, dx \ = \ -\tfrac{1}{2} e^{-nx^2}\Big|_0^1 \ = \ \tfrac{1}{2}(1 - e^{-n})$

Then $\displaystyle \lim_{n \to \infty} \int_0^1 s_n(x)\, dx \ = \ \lim_{n \to \infty} \tfrac{1}{2}(1 - e^{-n}) \ = \ \tfrac{1}{2}$

Also, $\displaystyle s(x) = \lim_{n \to \infty} s_n(x) = \lim_{n \to \infty} nxe^{-nx^2} = 0 \ $ whether $\ x = 0 \ $ or $\ 0 < x \leqq 1.$ Then

$$\int_0^1 s(x)\, dx \ = \ 0$$

It follows that $\ \displaystyle\lim_{n \to \infty} \int_0^1 s_n(x)\, dx \ \neq \ \int_0^1 \lim_{n \to \infty} s_n(x)\, dx,$ i.e. the limit cannot be taken under the integral sign.

(b) Although the sequence $s_n(x)$ converges to 0 in $[0, 1]$ it does not converge uniformly to 0 [see Problem 1.52, page 23]. If the convergence were uniform the results would have been equal. The fact that it is not uniform allows for the possibility [but not a certainty] that the results are not equal.

For a case where the sequence is not uniformly convergent but the integrals are nevertheless equal, see Problem A.59.

A.28. Let $\ s(x) = \displaystyle\sum_{n=1}^{\infty} \frac{\sin nx}{n^3}.$ Prove that $\ \displaystyle\int_0^\pi s(x)\, dx = 2\sum_{n=1}^{\infty} \frac{1}{(2n-1)^4}.$

We have $\ \left| \dfrac{\sin nx}{n^3} \right| \leqq \dfrac{1}{n^3}.$ Then since $\ \displaystyle\sum_{n=1}^{\infty} \frac{1}{n^3}$ converges, the series for $s(x)$ is uniformly convergent for all x and in particular for $\ 0 \leqq x \leqq \pi,$ by the Weierstrass M test [page 10]. Thus by Problem A.25 the series can be integrated term by term and we have

$$\int_0^\pi s(x)\, dx \ = \ \int_0^\pi \left(\sum_{n=1}^{\infty} \frac{\sin nx}{n^3} \right) dx \ = \ \sum_{n=1}^{\infty} \int_0^\pi \frac{\sin nx}{n^3}\, dx$$

$$= \ \sum_{n=1}^{\infty} \frac{1 - \cos n\pi}{n^4} \ = \ 2\left(\frac{1}{1^4} + \frac{1}{3^4} + \frac{1}{5^4} + \cdots \right)$$

$$= \ 2\sum_{n=1}^{\infty} \frac{1}{(2n-1)^4}$$

A.29. Give an example of a sequence of uniformly bounded functions $f_n(x)$ such that

$$\lim_{n \to \infty} \int_a^b f_n(x)\, dx$$

exists but

$$\int_a^b \lim_{n \to \infty} f_n(x)\, dx$$

does not exist.

Let $\ a = 0, \ b = 1 \ $ and consider the rational numbers in the interval $[0, 1].$ Since the rational numbers in $[0, 1]$ are denumerable [Problem 1.16, page 13], we can write them in the form of a sequence $r_1, r_2, r_3, \ldots.$

Define $\qquad\qquad f_n(x) \ = \ \begin{cases} 1 & \text{for } \ x = r_1, r_2, \ldots, r_n \\ 0 & \text{otherwise} \end{cases}$

Then $\qquad\qquad\qquad\qquad \displaystyle\int_0^1 f_n(x)\, dx \ = \ 0$

and so exists. However, since

$$\lim_{n \to \infty} f_n(x) \;=\; \begin{cases} 1 & \text{if } x \text{ is rational} \\ 0 & \text{if } x \text{ is irrational} \end{cases}$$

we see by Problem A.2 that

$$\int_0^1 \lim_{n \to \infty} f_n(x)\, dx$$

does not exist.

IMPROPER RIEMANN INTEGRALS

A.30. Prove that $\displaystyle\int_1^\infty \frac{dx}{(x+2)^3}$ exists and find its value.

We have by definition,

$$\int_1^\infty \frac{dx}{(x+2)^3} \;=\; \lim_{b \to \infty} \int_1^b \frac{dx}{(x+2)^3} \;=\; \lim_{b \to \infty} \frac{(x+2)^{-2}}{-2}\Big|_1^b$$

$$=\; \lim_{b \to \infty} \left(\frac{1}{18} - \frac{1}{2(b+2)^2} \right)$$

$$=\; \frac{1}{18}$$

A.31. Prove that $\displaystyle\int_{-7}^2 \frac{dx}{(x-1)^{5/3}}$ exists as an improper Riemann integral provided that it is taken in the Cauchy principal value sense.

The integrand $1/(x-1)^{5/3}$ is unbounded at $x=1$ so that the integral is improper. In the Cauchy principal value sense we define

$$\int_{-7}^2 \frac{dx}{(x-1)^{5/3}} \;=\; \lim_{\epsilon \to 0} \left[\int_{-7}^{1-\epsilon} \frac{dx}{(x-1)^{5/3}} + \int_{1+\epsilon}^2 \frac{dx}{(x-1)^{5/3}} \right]$$

$$=\; \lim_{\epsilon \to 0} \left[-\frac{3}{2}(x-1)^{-2/3}\Big|_{-7}^{1-\epsilon} - \frac{3}{2}(x-1)^{-2/3}\Big|_{1+\epsilon}^2 \right]$$

$$=\; \lim_{\epsilon \to 0} \left[-\frac{3}{2}(-\epsilon)^{-2/3} + \frac{3}{2}(-8)^{-2/3} - \frac{3}{2}(1)^{-2/3} + \frac{3}{2}(\epsilon)^{2/3} \right]$$

$$=\; -9/8$$

Note that if we define

$$\int_{-7}^2 \frac{dx}{(x-1)^{5/3}} \;=\; \lim_{\substack{\epsilon_1 \to 0 \\ \epsilon_2 \to 0}} \left[\int_{-7}^{1-\epsilon_1} \frac{dx}{(x-1)^{5/3}} + \int_{1+\epsilon_2}^2 \frac{dx}{(x-1)^{5/3}} \right]$$

where $\epsilon_1 \neq \epsilon_2$ the integral does not exist.

A.32. Prove that $\displaystyle\int_0^\infty \frac{\sin x}{x}\, dx$ exists.

The Riemann integral $\displaystyle\int_0^1 \frac{\sin x}{x}\, dx$

exists. Also, on integrating by parts we have

$$\int_1^b \frac{\sin x}{x}\, dx \;=\; -\frac{\cos x}{x}\Big|_1^b + \int_1^b \frac{\cos x}{x^2}\, dx \;=\; \cos 1 - \frac{\cos b}{b} + \int_1^b \frac{\cos x}{x^2}\, dx$$

Now $\displaystyle\left| \int_1^b \frac{\cos x}{x^2}\, dx \right| \;\leqq\; \int_1^b \left| \frac{\cos x}{x^2} \right| dx \;\leqq\; \int_1^b \frac{dx}{x^2} \;=\; 1 - \frac{1}{b}$

Then taking the limit as $b \to \infty$, we see that

$$\lim_{b \to \infty} \int_1^b \frac{\cos x}{x^2} \, dx = \int_1^\infty \frac{\cos x}{x^2} \, dx$$

exists as a Riemann integral. Now since

$$\int_0^b \frac{\sin x}{x} \, dx = \int_0^1 \frac{\sin x}{x} \, dx + \int_1^b \frac{\sin x}{x} \, dx$$

we see that $\quad \int_0^\infty \frac{\sin x}{x} \, dx = \lim_{b \to \infty} \int_0^b \frac{\sin x}{x} \, dx = \int_0^1 \frac{\sin x}{x} \, dx + \int_1^\infty \frac{\sin x}{x} \, dx$

exists.

A.33. Prove that $\displaystyle\int_0^\infty \left| \frac{\sin x}{x} \right| dx$ diverges, i.e. $\displaystyle\int_0^\infty \frac{\sin x}{x} \, dx$ is not absolutely convergent.

If n is any positive integer, we have

$$\int_0^{n\pi} \left| \frac{\sin x}{x} \right| dx = \int_0^\pi \left| \frac{\sin x}{x} \right| dx + \int_\pi^{2\pi} \left| \frac{\sin x}{x} \right| dx + \cdots + \int_{(n-1)\pi}^{n\pi} \left| \frac{\sin x}{x} \right| dx$$

$$= \int_0^\pi \left| \frac{\sin u}{u} \right| du + \int_0^\pi \left| \frac{\sin u}{u + \pi} \right| du + \cdots + \int_0^\pi \left| \frac{\sin u}{u + (n-1)\pi} \right| du$$

$$= \int_0^\pi \frac{\sin u}{u} \, du + \int_0^\pi \frac{\sin u}{u + \pi} \, du + \cdots + \int_0^\pi \frac{\sin u}{u + (n-1)\pi} \, du$$

$$> \frac{2}{\pi} + \frac{2}{2\pi} + \cdots + \frac{2}{(n-1)\pi}$$

$$= \frac{2}{\pi}\left(1 + \frac{1}{2} + \cdots + \frac{1}{n-1} \right)$$

But since the limit of the last series as $n \to \infty$ is infinite, the required result follows.

Supplementary Problems

DEFINITION OF THE RIEMANN INTEGRAL

A.34. Use the definition to show that $f(x) = 1$ is Riemann integrable in $[a, b]$ and find its value.

A.35. Use the definition of the Riemann integral to obtain the value of $\displaystyle\int_{-1}^1 f(x) \, dx$ where
$$f(x) = \begin{cases} -2, & -1 \leqq x < 0 \\ 5, & 0 < x \leqq 1 \end{cases}.$$

THEOREMS ON RIEMANN INTEGRATION

A.36. Prove Theorem A-11, page 156.

A.37. Prove that if $f(x)$ is Riemann integrable in $[a, b]$, then it is also Riemann integrable in any sub-interval $[c, d]$ of $[a, b]$.

A.38. Prove that the lower sums cannot decrease by adding points of subdivision [see remark at the end of Problem A.4].

A.39. Prove Theorem A-13, page 156.

A.40. Prove that $f(x)$ is Riemann integrable in $[a, b]$ if $f(x)$ is bounded and has (a) one point of discontinuity in $[a, b]$, (b) any finite number of points of discontinuity in $[a, b]$.

A.41. Prove that the Riemann integral of $f(x)$ does not change in value if the values of $f(x)$ in $[a, b]$ are changed at a finite number of points.

A.42. Prove Theorem A-18, page 157.

A.43. If $f(x)$ is continuous in $[a, b]$ and $f(x) \leq M$, then if

$$\int_a^b f(x)\, dx \;\geq\; M(b-a)$$

prove that $f(x) = M$ identically in $[a, b]$.

A.44. Obtain the following generalization of Theorem A-16, page 156: If $f(x)$ and $g(x)$ are Riemann integrable in $[a, b]$, $g(x) \geq 0$ and $m \leq f(x) \leq M$, then

$$m \int_a^b g(x)\, dx \;\leq\; \int_a^b f(x)\, g(x)\, dx \;\leq\; M \int_a^b g(x)\, dx$$

A.45. Use Problem A.44 to prove that if $f(x)$ is continuous in $[a, b]$ in addition to the other assumptions, then there is a number ξ such that

$$\int_a^b f(x)\, g(x)\, dx \;=\; f(\xi) \int_a^b g(x)\, dx$$

This is called the *generalized mean-value theorem* [see Theorem A-17, page 156].

A.46. If $f(x)$ is non-negative and continuous in $[a, b]$ and if $f(c) > 0$ for some value c in $[a, b]$, prove that

$$\int_a^b f(x)\, dx \;>\; 0$$

A.47. If $f(x)$ is continuous in $[a, b]$ and

$$\int_a^b |f(x)|^2\, dx \;=\; 0$$

prove that $f(x) = 0$ in $[a, b]$.

A.48. Prove Theorem A-12, page 156. [*Hint*: First assume that $f(x)$ and $g(x)$ are both positive and have upper and lower bounds $M_k^{(1)}, m_k^{(1)}$ and $M_k^{(2)}, m_k^{(2)}$ respectively in $[x_{k-1}, x_k]$. Then prove that if M_k, m_k are upper and lower bounds of $f(x)\, g(x)$ in $[x_{k-1}, x_k]$, then $M_k - m_k \leq M_k^{(1)}(M_k^{(2)} - m_k^{(2)}) + m_k^{(2)}(M_k^{(1)} - m_k^{(1)})$ and thus show that $S - s < \epsilon$. If $f(x)$ and $g(x)$ are not both positive, add suitable constants so as to make them positive.]

A.49. Prove that the definition of the Riemann integral given on page 154 follows from the definition as a limit of a sum.

DIFFERENTIATION AND INTEGRATION

A.50. Let $f(x) = \begin{cases} x, & 0 \leq x < 2 \\ 4, & 2 < x \leq 5 \end{cases}$. (a) Evaluate $F(x) = \int_0^x f(u)\, du$, $0 \leq x \leq 5$. (b) Show that $F(x)$ is continuous in $0 \leq x \leq 5$. (c) Does $F'(x) = f(x)$? Explain. (d) Does $\int_0^5 F'(x)\, dx = F(5) - F(0)$? Explain.

A.51. Consider the functions (a) $f(x) = x^3 \cos(1/x)$, $0 \leq x \leq 1/\pi$, (b) $f(x) = \begin{cases} x^3 \cos 1/x, & 0 < x \leq 1/\pi \\ 0, & x = 0 \end{cases}$. Determine whether

$$\int_0^\pi f'(x)\, dx$$

exists and if it does find its value.

A.52. Prove the *second mean-value theorem*: If $f(x)$ and $g(x)$ are Riemann integrable in $[a, b]$ and $f(x)$ is monotonic in $[a, b]$, then for some value η such that $a \leq \eta \leq b$

$$\int_a^b f(x)\, g(x)\, dx \;=\; f(a) \int_a^\eta g(x)\, dx \;+\; f(b) \int_\eta^b g(x)\, dx$$

[*Hint*: Let $G(x) = \int_a^x g(t)\, dt$. Write $\int_a^b f(x)\, g(x)\, dx = \int_a^b f(x)\, G'(x)\, dx$ and then use integration by parts.]

A.53. Prove Theorem A-24, page 157.

A.54. Use the definition of the Riemann integral as a limit of a sum to show that

$$\lim_{n \to \infty} \left(\frac{n}{1^2 + n^2} + \frac{n}{2^2 + n^2} + \cdots + \frac{n}{n^2 + n^2} \right) \;=\; \frac{\pi}{4}$$

THEOREMS ON SEQUENCES AND SERIES

A.55. Prove Theorem A-25, page 158.

A.56. Prove that $\displaystyle\int_0^\pi \left(\frac{\cos 2x}{1 \cdot 3} + \frac{\cos 4x}{3 \cdot 5} + \frac{\cos 6x}{5 \cdot 7} + \cdots \right) dx \;=\; 0$.

A.57. Suppose that the sequence of functions $\langle f_n(x) \rangle$ converges to $f(x)$ in $[a, b]$. State and prove a theorem giving sufficient conditions under which the sequence of derivatives $\langle f'_n(x) \rangle$ converges to $f'(x)$.

A.58. Prove that $F(x) = \displaystyle\sum_{n=1}^\infty \frac{\sin nx}{\sinh n\pi}$ is continuous and has continuous derivatives of all orders for all values of x.

A.59. Let $s_n(x) = \dfrac{nx}{1 + n^2 x^2}$ for $0 \leq x \leq 1$. (a) Prove that $\langle s_n(x) \rangle$ is not uniformly convergent. (b) Show that

$$\lim_{n \to \infty} \int_0^1 s_n(x)\, dx \;=\; \int_0^1 \lim_{n \to \infty} s_n(x)\, dx$$

and explain.

A.60. Use Problem A.59 to give an example of a series which can be integrated term by term but which is not uniformly convergent.

A.61. Let $s_n(x) = \begin{cases} nx, & 0 \leq x \leq 1/n \\ 1, & 1/n < x \leq 1 \end{cases}$. Determine whether

$$\lim_{n \to \infty} \int_0^1 s_n(x)\, dx \;=\; \int_0^1 \lim_{n \to \infty} s_n(x)\, dx$$

and explain.

IMPROPER RIEMANN INTEGRALS

A.62. Investigate the existence of the Riemann integrals (a) $\int_{-1}^{1} \frac{dx}{\sqrt{x+1}}$, (b) $\int_{0}^{4} \frac{dx}{(x-1)^2}$ and find their values if they exist.   *Ans.* (a) $2\sqrt{2}$, (b) does not exist

A.63. Evaluate the following integrals if they exist. (a) $\int_{0}^{\infty} \frac{dx}{(x+1)^4}$, (b) $\int_{-\infty}^{\infty} \frac{dx}{x^2+1}$, (c) $\int_{1}^{\infty} \frac{dx}{\sqrt{x}}$.
Ans. (a) $\frac{1}{3}$, (b) π, (c) does not exist

A.64. Prove that if $0 \leqq f(x) \leqq g(x)$ and $g(x)$ has an improper Riemann integral in $[a, b]$ which exists, then so also does $f(x)$.

A.65. Use the result of Problem A.64 to show that $\int_{0}^{1} \frac{dx}{\sqrt{x} + \sin^2 x}$ exists.

A.66. Show that the integrals (a) $\int_{-\infty}^{\infty} \frac{dx}{(x-2)^5}$, (b) $\int_{0}^{2} \frac{dx}{x^2-1}$ exist only in the Cauchy principal value sense and find these values.   *Ans.* (a) 0, (b) $-\frac{1}{2} \ln 3$

Summability of Fourier Series

CONVERGENCE IN THE CESARO SENSE

Let $s_n = u_1 + u_2 + \cdots + u_n$ be the nth partial sum [i.e. the sum of the first n terms] of the series $u_1 + u_2 + \cdots$. If the sequence of partial sums $s_1, s_2, s_3, \ldots$ converges to the limit l, then by definition the series converges to l or has sum l.

Suppose however that the series does not converge. Then consider a new sequence

$$s_1, \quad \frac{s_1 + s_2}{2}, \quad \frac{s_1 + s_2 + s_3}{3}, \quad \ldots \tag{1}$$

where the nth term is the arithmetic mean of the first n partial sums. If this sequence converges to p, we say that the series $u_1 + u_2 + \cdots$ or $\sum u_k$ is *summable in the Cesaro sense*, or is *C-1 summable*, to p.

In case the series $\sum u_k$ does converge to l, then the series is also C-1 summable to l [see Problem B.3].

The process can be repeated if necessary and we speak of C-2 summability, C-3 summability, etc.

CESARO SUMMABILITY OF FOURIER SERIES. FEJER'S THEOREM

The Cesaro summability of Fourier series was investigated by *Fejer*. Since the partial sums of a Fourier series are given by

$$s_n(x) = \frac{1}{2\pi} \int_0^\pi [f(x+u) + f(x-u)] \frac{\sin(n + \frac{1}{2})u}{\sin \frac{1}{2}u} \, du \tag{2}$$

where $n = 0, 1, 2, \ldots$ [see Problem 8.8, page 139], we obtain the nth term of the arithmetic means [see Problem B.4],

$$t_n(x) = \frac{s_0(x) + \cdots + s_{n-1}(x)}{n} = \frac{1}{2n\pi} \int_0^\pi [f(x+u) + f(x-u)] \frac{\sin^2 \frac{1}{2}nu}{\sin^2 \frac{1}{2}u} \, du \tag{3}$$

We can then show that for some function $t(x)$,

$$t_n(x) - t(x) = \frac{1}{2n\pi} \int_0^\pi [f(x+u) + f(x-u) - 2t(x)] \frac{\sin^2 \frac{1}{2}nu}{\sin^2 \frac{1}{2}u} \, du \tag{4}$$

and can prove the following theorems.

Theorem B-1 **[Fejer].** If $f(x+0)$ and $f(x-0)$ exist, then the Fourier series for $f(x)$ converges in the Cesaro sense to $\frac{1}{2}[f(x+0) + f(x-0)]$, i.e. the Fourier series for $f(x)$ is C-1 summable to $\frac{1}{2}[f(x+0) + f(x-0)]$.

Theorem B-2. The Fourier series for $f(x)$ is C-1 summable to $f(x)$ almost everywhere.

Solved Problems

CONVERGENCE IN THE CESARO SENSE

B.1. Prove that the series $1 - 1 + 1 - 1 + \cdots$ does not converge.

If s_n is the sum of the first n terms, we have $s_1 = 1$, $s_2 = 0$, $s_3 = 1$, $s_4 = 0$, ... Thus the sequence of partial sums is $1, 0, 1, 0, \ldots$ which does not converge.

B.2. Prove that the series of Problem B.1 is C-1 summable to $\frac{1}{2}$.

From Problem B.1 we see that the sequence of arithmetic means of the partial sums is

$$s_1, \quad \frac{s_1 + s_2}{2}, \quad \frac{s_1 + s_2 + s_3}{3}, \quad \frac{s_1 + s_2 + s_3 + s_4}{4}, \quad \cdots$$

or

$$1, \quad \frac{1 + 0}{2}, \quad \frac{1 + 0 + 1}{3}, \quad \frac{1 + 0 + 1 + 0}{4}, \quad \cdots$$

i.e.

$$1, \quad \tfrac{1}{2}, \quad \tfrac{2}{3}, \quad \tfrac{1}{2}, \quad \tfrac{3}{5}, \quad \tfrac{1}{2}, \quad \cdots$$

If we let t_n be the nth term of the last sequence, we see that

$$t_n = \begin{cases} \dfrac{n+1}{2n} & n = 1, 3, 5, \ldots \\ \dfrac{1}{2} & n = 2, 4, 6, \ldots \end{cases}$$

Then since $\lim\limits_{n \to \infty} t_n = \frac{1}{2}$, it follows that the series $1 - 1 + 1 - 1 + \cdots$ is C-1 summable to $\frac{1}{2}$.

B.3. (a) If $\lim\limits_{n \to \infty} u_n = l$, prove that $\lim\limits_{n \to \infty} \dfrac{u_1 + u_2 + \cdots + u_n}{n} = l$ and (b) discuss the significance of this in connection with Cesaro summability.

(a) Let $u_n = v_n + l$. Then we must show that

$$\lim_{n \to \infty} \frac{v_1 + v_2 + \cdots + v_n}{n} = 0 \quad \text{if} \quad \lim_{n \to \infty} v_n = 0$$

Now if $n > p$,

$$\frac{v_1 + v_2 + \cdots + v_n}{n} = \frac{v_1 + v_2 + \cdots + v_p}{n} + \frac{v_{p+1} + v_{p+2} + \cdots + v_n}{n}$$

so that

$$\left| \frac{v_1 + v_2 + \cdots + v_n}{n} \right| \leqq \frac{|v_1 + v_2 + \cdots + v_p|}{n} + \frac{|v_{p+1}| + |v_{p+2}| + \cdots + |v_n|}{n} \tag{1}$$

Since $\lim\limits_{n \to \infty} v_n = 0$, we can choose p so that given any $\epsilon > 0$, $|v_n| < \epsilon/2$ for $n > p$. Then

$$\frac{|v_{p+1}| + |v_{p+2}| + \cdots + |v_n|}{n} < \frac{\epsilon/2 + \epsilon/2 + \cdots + \epsilon/2}{n} = \frac{(n-p)\epsilon/2}{n} < \frac{\epsilon}{2} \tag{2}$$

After choosing p we can choose n_0 so that for $n > n_0 > p$,

$$\frac{|v_1 + v_2 + \cdots + v_p|}{n} < \frac{\epsilon}{2} \tag{3}$$

Then using (2) and (3), (1) becomes

$$\left| \frac{v_1 + v_2 + \cdots + v_n}{n} \right| < \frac{\epsilon}{2} + \frac{\epsilon}{2} = \epsilon \quad \text{for} \quad n > n_0$$

thus proving the required result.

(b) The result shows that if the sequence of partial sums of a series does converge to l, i.e. if the series converges to l, then the series is C-1 summable to l.

CESARO SUMMABILITY OF FOURIER SERIES.   FEJER'S THEOREM

B.4.   Using Problem 8.8, page 139, for the partial sums $s_n(x)$ of a Fourier series, prove that

$$t_n(x) \;=\; \frac{s_0(x) + s_1(x) + \cdots + s_{n-1}(x)}{n} \;=\; \frac{1}{2n\pi} \int_0^\pi [f(x+u) + f(x-u)] \frac{\sin^2 \frac{1}{2}nu}{\sin^2 \frac{1}{2}u}\, du$$

From Problem 8.8, we have

$$s_n(x) \;=\; \frac{1}{2\pi} \int_0^\pi [f(x+u) + f(x-u)] \frac{\sin(n+\frac{1}{2})u}{\sin \frac{1}{2}u}\, du$$

for $n = 0, 1, 2, \ldots$ so that

$$t_n(x) \;=\; \frac{s_0(x) + s_1(x) + \cdots + s_{n-1}(x)}{n}$$

$$=\; \frac{1}{2n\pi} \int_0^\pi [f(x+u) + f(x-u)] \left[\frac{\sin \frac{1}{2}u + \sin \frac{3}{2}u + \cdots + \sin(n - \frac{1}{2})u}{\sin \frac{1}{2}u} \right] du$$

Now consider $f_n \;=\; \sin \frac{1}{2}u + \sin \frac{3}{2}u + \cdots + \sin(n - \frac{1}{2})u$

Multiplying by $\sin \frac{1}{2}u$,

$$f_n \sin \frac{1}{2}u \;=\; \sin^2 \frac{1}{2}u + \sin \frac{3}{2}u \sin \frac{1}{2}u + \cdots + \sin(n - \frac{1}{2})u \sin \frac{1}{2}u$$

$$=\; \tfrac{1}{2}(1 - \cos u) + \tfrac{1}{2}(\cos u - \cos 2u) + \cdots + \tfrac{1}{2}(\cos(n-1)u - \cos nu)$$

$$=\; \tfrac{1}{2}(1 - \cos nu) \;=\; \sin^2 \tfrac{1}{2}nu$$

Thus $f_n \;=\; \dfrac{\sin^2 \frac{1}{2}nu}{\sin \frac{1}{2}u}$ (1)

and we obtain as required,

$$t_n(x) \;=\; \frac{1}{2n\pi} \int_0^\pi [f(x+u) + f(x-u)] \frac{\sin^2 \frac{1}{2}nu}{\sin^2 \frac{1}{2}u}\, du \qquad (2)$$

B.5.   (a) Prove that

$$\frac{1}{2n\pi} \int_0^\pi \frac{2 \sin^2 \frac{1}{2}nu}{\sin^2 \frac{1}{2}u}\, du \;=\; 1$$

and (b) thus show that if $t(x)$ is some function of x

$$t_n(x) \,-\, t(x) \;=\; \frac{1}{2n\pi} \int_0^\pi [f(x+u) + f(x-u) - 2t(x)] \frac{\sin^2 \frac{1}{2}nu}{\sin^2 \frac{1}{2}u}\, du$$

(a) If $f(x) = 1$, then $t_n(x) = 1$ also. Thus from (2) of Problem B.4 we obtain as required,

$$1 \;=\; \frac{1}{2n\pi} \int_0^\pi \frac{2 \sin^2 \frac{1}{2}nu}{\sin^2 \frac{1}{2}u}\, du \qquad (1)$$

(b) Multiplying both sides of (1) by $t(x)$ and subtracting from equation (2) of Problem B.4, we obtain

$$t_n(x) \,-\, t(x) \;=\; \frac{1}{2n\pi} \int_0^\pi [f(x+u) + f(x-u)] \frac{\sin^2 \frac{1}{2}nu}{\sin^2 \frac{1}{2}u}\, du$$

$$-\; \frac{1}{2n\pi} \int_0^\pi 2t(x) \frac{\sin^2 \frac{1}{2}nu}{\sin^2 \frac{1}{2}u}\, du$$

$$=\; \frac{1}{2n\pi} \int_0^\pi [f(x+u) + f(x-u) - 2t(x)] \frac{\sin^2 \frac{1}{2}nu}{\sin^2 \frac{1}{2}u}\, du$$

B.6. Prove that a Fourier series is C-1 summable to $t(x)$ if and only if for any number δ such that $0 < \delta < \pi$,

$$\lim_{n \to \infty} \frac{1}{n} \int_0^\delta F(u) \frac{\sin^2 \frac{1}{2}nu}{\sin^2 \frac{1}{2}u} \, du \;=\; 0$$

where $F(u) = f(x+u) + f(x-u) - 2t(x)$.

From Problem B.3(b) it is clear that a Fourier series is C-1 summable to $t(x)$ if

$$\lim_{n \to \infty} \frac{1}{n} \int_0^\pi F(u) \frac{\sin^2 \frac{1}{2}nu}{\sin^2 \frac{1}{2}u} \, du \;=\; 0$$

i.e.

$$\lim_{n \to \infty} \left\{ \frac{1}{n} \int_0^\delta F(u) \frac{\sin^2 \frac{1}{2}nu}{\sin^2 \frac{1}{2}u} \, du + \frac{1}{n} \int_\delta^\pi F(u) \frac{\sin^2 \frac{1}{2}nu}{\sin^2 \frac{1}{2}u} \, du \right\} \;=\; 0$$

But since

$$\left| \frac{1}{n} \int_\delta^\pi F(u) \frac{\sin^2 \frac{1}{2}nu}{\sin^2 \frac{1}{2}u} \, du \right| \;\leqq\; \frac{1}{n} \int_\delta^\pi \frac{|F(u)|}{\sin^2 \frac{1}{2}u} \, du$$

we see that for $0 < \delta < \pi$,

$$\lim_{n \to \infty} \frac{1}{n} \int_\delta^\pi F(u) \frac{\sin^2 \frac{1}{2}nu}{\sin^2 \frac{1}{2}u} \, du \;=\; 0$$

Thus we see that the Fourier series is C-1 summable to $t(x)$ if and only if

$$\lim_{n \to \infty} \frac{1}{n} \int_0^\delta F(u) \frac{\sin^2 \frac{1}{2}nu}{\sin^2 \frac{1}{2}u} \, du \;=\; 0$$

B.7. Prove that a Fourier series is C-1 summable to $t(x)$ if and only if

$$\lim_{n \to \infty} \frac{1}{n} \int_0^\delta F(u) \frac{\sin^2 \frac{1}{2}nu}{u^2} \, du \;=\; 0$$

We have

$$\left| \frac{1}{n} \int_0^\delta F(u) \left[\frac{\sin^2 \frac{1}{2}nu}{\sin^2 \frac{1}{2}u} - \frac{\sin^2 \frac{1}{2}nu}{(\frac{1}{2}u)^2} \right] du \right|$$

$$= \; \left| \frac{1}{n} \int_0^\delta F(u) \sin^2 \tfrac{1}{2}nu \left[\frac{1}{\sin^2 \frac{1}{2}u} - \frac{1}{(\frac{1}{2}u)^2} \right] du \right|$$

$$\leqq \; \frac{1}{n} \int_0^\delta |F(u)| \left[\frac{1}{\sin^2 \frac{1}{2}u} - \frac{1}{(\frac{1}{2}u)^2} \right] du$$

since

$$\frac{1}{\sin^2 \frac{1}{2}u} - \frac{1}{\frac{1}{2}u^2}$$

is non-negative and continuous for $0 < u \leqq \delta$ and has a finite limit as $u \to 0$.

Then using Problem B.6

$$\lim_{n \to \infty} \frac{1}{n} \int_0^\delta F(u) \frac{\sin^2 \frac{1}{2}nu}{(\frac{1}{2}u)^2} \, du \;=\; 0$$

which yields the required result.

B.8. Prove Fejer's theorem.

Let $t(x) = \frac{1}{2}[f(x+0) + f(x-0)]$. Then

$$F(u) \;=\; f(x+u) + f(x-u) - [f(x+0) + f(x-0)]$$

By definition, $\lim\limits_{u \to 0+} f(x + u) = f(x + 0)$, $\lim\limits_{u \to 0+} f(x - u) = f(x - 0)$ and we have $\lim\limits_{u \to 0+} F(u) = 0$. Thus, given $\epsilon > 0$, we can find $\delta_1 > 0$ so that $|F(u)| < \epsilon$ whenever $|u| < \delta_1$ or $u < \delta_1$, since u is non-negative.

Then we have for fixed $\delta_1 < \pi$,

$$\left| \frac{1}{n} \int_0^\delta F(u) \frac{\sin^2 \frac{1}{2}nu}{u^2}\, du \right| \leqq \left| \frac{1}{n} \int_0^{\delta_1} F(u) \frac{\sin^2 \frac{1}{2}nu}{u^2}\, du \right| + \left| \frac{1}{n} \int_{\delta_1}^\delta F(u) \frac{\sin^2 \frac{1}{2}nu}{u^2}\, du \right|$$

$$\leqq \frac{1}{n} \int_0^{\delta_1} |F(u)| \frac{\sin^2 \frac{1}{2}nu}{u^2}\, du + \frac{1}{n} \int_{\delta_1}^\delta |F(u)| \frac{\sin^2 \frac{1}{2}nu}{u^2}\, du$$

$$\leqq \frac{\epsilon}{n} \int_0^{\delta_1} \frac{\sin^2 \frac{1}{2}nu}{u^2}\, du + \frac{1}{n} \int_{\delta_1}^\delta \frac{|F(u)|}{u^2}\, du$$

$$\leqq \frac{\epsilon}{2} \int_0^{n\delta_1/2} \frac{\sin^2 v}{v^2}\, dv + \frac{1}{n} \int_{\delta_1}^\delta \frac{|F(u)|}{\delta_1^2}\, du$$

$$\leqq \frac{\epsilon}{2} \int_0^\infty \frac{\sin^2 v}{v^2}\, dv + \frac{1}{n\delta_1^2} \int_{\delta_1}^\delta |F(u)|\, du$$

$$= C\epsilon + \frac{1}{n\delta_1^2} \int_{\delta_1}^\delta |F(u)|\, du$$

Now since δ_1 is fixed, $\qquad \lim\limits_{n \to \infty} \frac{1}{n\delta_1^2} \int_{\delta_1}^\delta |F(u)|\, du = 0$

Thus since ϵ can be taken as arbitrarily small, we have

$$\lim\limits_{n \to \infty} \frac{1}{n} \int_0^\delta F(u) \frac{\sin^2 \frac{1}{2}nu}{u^2}\, du = 0 \qquad\qquad (1)$$

But (1) is a necessary and sufficient condition that the Fourier series be C-1 summable to $t(x) = \frac{1}{2}[f(x + 0) + f(x - 0)]$, and so Fejer's theorem is proved.

Appendix C

Double Lebesgue Integrals and Fubini's Theorem

LEBESGUE MEASURE IN THE PLANE

The theory of Lebesgue measure and integration in the plane R^2 is in many ways analogous to that for the real line R. Thus instead of an interval I and length of an interval $L(I)$ on the real line, we have a rectangle P [which can be considered as the Cartesian product of two intervals I_1 and I_2] and area of a rectangle $A(P)$ [defined as the product of the lengths of I_1 and I_2]. The rectangle P is *closed* if I_1 and I_2 are closed and *open* if I_1 and I_2 are open.

The measure of a rectangle is defined to be its area. As in the case of the real line we can prove that an open set in the plane can be expressed as a countable union of disjoint open rectangles. We then define the measure of an open set as the sum of the measures of these associated rectangles [which can be called *components*]. We can then define the *exterior or outer measure* of a planar set S, denoted by $m_e(S)$, as the least upper bound of the measures of all open sets containing S. Finally we say that S is *measurable* if for any set T,

$$m_e(T) = m_e(T \cap S) + m_e(T \cap \tilde{S}) \tag{1}$$

as in the one dimensional case.

The various theorems on measure in R^2 are in general analogous to those for R^1.

MEASURABLE FUNCTIONS IN THE PLANE

Let f be a function defined from a set of points (x, y) in the plane to the set of real numbers, so that f is a real function whose values can be designated by $f(x, y)$. We say that f is *measurable* if for any constant κ the set

$$\{(x, y) : f(x, y) > \kappa\} \tag{2}$$

is measurable. The various properties of measurable functions are analogous to those developed in Chapter 3.

The following theorem is important.

Theorem C-1. If $f(x, y)$ is measurable on a set $[(x, y) : a \leq x \leq b, c \leq y \leq d]$, then $f(x, y)$ is a measurable function of y for all $x \in [a, b]$ and $f(x, y)$ is a measurable function of x for all $y \in [c, d]$.

THE LEBESGUE INTEGRAL IN THE PLANE

Given a rectangle $P = \{(x, y) : a \leqq x \leqq b, \ c \leqq y \leqq d\}$, we can subdivide it into parts by using subdivision points

$$a = x_0 < x_1 < x_2 < \cdots < x_n = b, \qquad c = y_0 < y_1 < y_2 < \cdots < y_n = d$$

We can then define *upper* and *lower Lebesgue integrals* for bounded measurable functions $f(x, y)$ in exactly the same way as we defined them for bounded measurable functions $f(x)$ on the real line.

These are denoted respectively by

$$\overline{\iint_P} f(x, y) \, dx \, dy \quad \text{and} \quad \underline{\iint_P} f(x, y) \, dx \, dy \tag{3}$$

If these are equal the common value is denoted by

$$\iint_P f(x, y) \, dx \, dy \tag{4}$$

and is called the *Lebesgue double integral* and we say that f or $f(x, y)$ is *Lebesgue integrable* on P.

The result can be extended to any set E in R^2 and in such case the double integral is denoted by

$$\iint_E f(x, y) \, dx \, dy \tag{5}$$

FUBINI'S THEOREM

Let $f(x, y)$ be Lebesgue integrable on P. Then $f(x, y)$ is Lebesgue integrable over $[a, b]$ for almost all y such that $c \leqq y \leqq d$, i.e. the integral

$$\int_a^b f(x, y) \, dx \tag{6}$$

exists for almost all y such that $c \leqq y \leqq d$.

Furthermore the integral (6) is Lebesgue integrable on $[c, d]$ so that

$$\int_c^d \left[\int_a^b f(x, y) \, dx \right] dy \tag{7}$$

exists.

In a similar manner we can demonstrate the existence of

$$\int_a^b \left[\int_c^d f(x, y) \, dy \right] dx \tag{8}$$

The integrals (7) and (8), where integration is performed first with respect to one variable and then with respect to the other, are often called *iterated integrals*.

The fundamental theorem connecting double integrals and iterated integrals is called *Fubini's theorem*.

Theorem C-2 [**Fubini**].   If $f(x, y)$ is measurable on the set

$$P = \{(x, y) : a \leqq x \leqq b, \ c \leqq y \leqq d\}$$

then

$$\iint_P f(x, y) \, dx \, dy = \int_c^d \left[\int_a^b f(x, y) \, dx \right] dy = \int_a^b \left[\int_c^d f(x, y) \, dy \right] dx \tag{9}$$

THE FUBINI-TONELLI-HOBSON THEOREM

Sometimes one of the iterated integrals is given and it is required to interchange the order of integration. The fundamental theorem on interchanging the order of integration is called the *Fubini-Tonelli-Hobson theorem*.

Theorem C-3 [**Fubini-Tonelli-Hobson**]. We have

$$\int_c^d \left[\int_a^b f(x,y)\,dx \right] dy \;=\; \int_a^b \left[\int_c^d f(x,y)\,dy \right] dx \tag{10}$$

provided that either one of the iterated integrals in (*10*) exists.

Since a function which is Lebesgue integrable is absolutely integrable, it follows that the order of integration can be interchanged if either of the sides in (*10*) is absolutely convergent.

The results can be extended to the case where the limits of integration are infinite.

MULTIPLE LEBESGUE INTEGRALS

The ideas presented above can be generalized to higher spaces such as R^n where $n > 2$. Thus for example if $n = 3$, i.e. three dimensional Euclidean space, we begin with the definition of the measure of a rectangular parallelepiped as its *volume* [the product of the lengths of three mutually perpendicular sides]. This leads to the notion of a *triple integral*. For $n > 3$ the ideas can be developed in an entirely analogous manner.

Index of Special Symbols
and Notations

The following list shows special symbols and notations used in this book together with the number of the page on which they first appear. Cases where a symbol has more than one meaning will be clear from the context.

Symbols

$A, B, \ldots$	sets, 1
$a, b, \ldots$	elements or members of sets, 1
a_n, b_n	Fourier coefficients, 130
$\aleph_0$	aleph null, cardinal number of a denumerable set, 5
$\aleph_1$ or c	aleph one or c, cardinal number of the set of real numbers R, 5
C	closed set, 30
$\mathcal{C}$	open covering, 7
c_k	generalized Fourier coefficients, 132
δ	norm of a partition, 154
δ, ϵ	positive numbers, 6, 7
f	function or mapping, 4
$\emptyset$	empty or null set, 1
$H_n(x)$	Hermite polynomials, 145
I	interval, 29
I	g.l.b. of upper sums for Lebesgue integrals, 53 [or Riemann integrals, 154]
$I_1, I_2, \ldots$	intervals, 9
$\mathcal{I}$	class of intervals, 34
J	l.u.b. of lower sums for Lebesgue intgerals, 53 [or Riemann integrals, 154]
J	open subcovering, 7
K	Cantor set, 5
l	limit of $f(x)$ as $x \to a$, 7
$L_n(x)$	Laguerre polynomials, 152
M	upper bound of $f(x)$, 154
m	lower bound of $f(x)$, 154
N	set of natural numbers or positive integers $1, 2, 3, \ldots, 2$
O	open set, 29
P	partition, 154
P	rectangle, 180
$P_n(x)$	Legendre polynomials, 152
Q	set of rational numbers, 2
R	set of real numbers, 2
$r_n(x)$	remainder of a series after n terms, 23
S	upper sum for Lebesgue integral, 53 [or Riemann integral, 154]

Notations

$A \sim B$	A is equivalent to B, 5
$(x_1, \ldots, x_n)$	ordered n-tuplet, 6
$d(x, y)$	distance between x and y, 6
$\|a\|$	absolute value of a, 6
R^n	Euclidean space of n dimensions, 6
S'	derived set of S, 7
$\bar{S}$	closure of S, 7
$\lim_{x \to a} f(x)$	limit of $f(x)$ as x approaches a, 7
$\lim_{x \to a+} f(x)$	right hand limit of $f(x)$ as x approaches a, 8
$\lim_{x \to a-} f(x)$	left hand limit of $f(x)$ as x approaches a, 8
$\langle a_n \rangle$	sequence $a_1, a_2, a_3, \ldots$, 8
$\lim_{n \to \infty} a_n$ or $\lim a_n$	limit of sequence $\langle a_n \rangle$ as $n \to \infty$, 8
$\bar{l}, \overline{\lim} a_n$ or $\lim \sup a_n$	limit superior, greatest limit or upper limit of $\langle a_n \rangle$, 9
$\underline{l}, \underline{\lim} a_n$ or $\lim \inf a_n$	limit inferior, least limit or lower limit of $\langle a_n \rangle$, 9
$\langle f_n \rangle$ or $\langle f_n(x) \rangle$	sequence of functions, 9
$\sum_{n=1}^{\infty} a_n$ or $\sum a_n$	series $a_1 + a_2 + \cdots$, 9
$L(I)$	length of an interval I, 29
$m(E)$	measure of a set E, 30
$m_e(E)$	exterior or outer measure of a set E, 30
$m_i(E)$	interior or inner measure of a set E, 32
$E\{f(x) > \kappa\}$	set of values $x \in E$ for which $f(x) > \kappa$, 43
$\max \{f_1(x), f_2(x)\}$	maximum of $f_1(x)$ and $f_2(x)$, 44
$\min \{f_1(x), f_2(x)\}$	minimum of $f_1(x)$ and $f_2(x)$, 44
$\chi(x)$	characteristic function, 49
$\overline{\int_a^b} f(x)\,dx$	upper Lebesgue integral [except in Appendix A where it is the upper Riemann integral], 54
$\underline{\int_a^b} f(x)\,dx$	lower Lebesgue integral [except in Appendix A where it is the lower Riemann integral], 54
$\int_a^b f(x)\,dx$	Lebesgue integral of $f(x)$ from a to b [except in Appendix A where it is the Riemann integral], 54
$(\mathscr{R}) \int_a^b f(x)\,dx$	Riemann integral of $f(x)$ from a to b, 54
$\int_a^b f(x)\,dx < \infty$	existence of the integral of $f(x)$ from a to b, 54
$\int_E f(x)\,dx$	Lebesgue integral of $f(x)$ on the set E, 55
E^+	$E[f(x) \geqq 0]$, 65
E^-	$E[f(x) < 0]$, 65
$[f(x)]_p$	$\begin{cases} f(x) & \text{for all } x \in E \text{ such that } f(x) \leqq p \\ p & \text{for all } x \in E \text{ such that } f(x) > p \end{cases}$, 72

INDEX